Taboo Knowledge
What Leaders of Science, Religion, & Politics Don't Want You to Know

By

Stephen Hawley Martin

WWW.OAKLEAPRESS.COM

Contents

Chapter One: The Truth Revealed

It's about saving face, and it's about staying in power. No one, particularly someone in power, wants to admit he or she is wrong. To do so is to lose credibility, and to lose credibility is to lose the confidence of those you lead. Suddenly, you'd not only be out of work, you'd be a has-been and a laughing-stock. And so, when it comes to something really, really big that you now realize you're wrong about, it becomes a matter of keeping word from getting out, and thereby maintaining power. That's why leaders of science, religion, and politics do not want you to know the truth about who and what you actually are. Letting word get out would totally shatter their worlds, and it would not only take power away from them, as you will see, it would put immense power in your hands.

No wonder they don't want you to know the truth.

The lion's share of the blame for the current situation falls on the leaders of the scientific community—the editors and publishers of scientific journals, science writers for national publications, and the department heads and leading research scientists in academia. In order to maintain their revered status, they continue to perpetuate a myth—one the majority of western scientists once really did think was true, but which in recent years has become increasingly difficult—impossible, once the truth gets out—to defend. It may be that some still believe it, but those who do either haven't given it serious thought, or they are in denial—because it does not hold up even to the most casual scrutiny.

On the other hand, many political leaders may actually believe what scientists continue to tell them, although some are clearly complicit in the charade. Faith leaders have their own reasons—strong ones—for denying or obfuscating the truth.

Who and What You Are Not

Let's cut to the chase. Who and what are you? The truth will be shocking to some, and so before we come to it, let us agree on who and what you are not.

Would you agree you that it is impossible to be something you can observe? For example, you are not the tree you can see across the street. You are not the house or apartment in which you live. You are not a penny on a sidewalk.

What you can observe includes more than objects. For example, although they may give you a sense of personal identity, you are not your job, you are not the country in which you were born, and you are not your religion. You are also not the color of your skin, and you are not the school or university you went to.

You are not your hands or your feet, and you are not your body.

"Wait. I'm not my body?"

That statement may raise doubt in you, but you are not something you can observe, and you can observe your body. In fact, the two words, "your body," suggest your body is not you, but rather, something that belongs to you. You can control the actions of your body. For example, you might decide to force it to dig a ditch, and if you do, after a few hours, you might say, "My body aches." And so it's not you that aches—it's your body—your back and your arms.

How about your mind? Do you think you are your mind?

You are not your mind, and, as you will see, realizing you are not your mind will empower you. Consider this. You can observe your thoughts as they come and they go. You may even observe them to a fault when they keep you awake at night. But you, the real you, do not have to be at the mercy of your thoughts, and therefore, your mind. Like Scarlett O'Hara, you can decide to think about tomorrow whatever problem has arisen in your mind today.

It's a fact. Anything you can—sometimes figuratively speaking—stand back and observe cannot possibly be you. Moreover, whether you realized it or not before now—whether you have ever exercised this prerogative or not—you can decide which thoughts to pay attention to and which ones to discard. You can decide what to think about, or not to think about. In other words, you

7

can control the actions of your mind, just as you can control the actions of your body, and this means your mind cannot possibly be you.

You may be surprised to learn who and what you are, and you may initially experience an impulse to reject it, so before it's revealed, consider this: There was a time when people thought the world was flat and that the earth was at the center of the universe—that the sun and the stars revolved around the earth. Now, just about everyone knows the earth is round, that the sun is at the center of the solar system, and that our solar system is one of trillions upon trillions.

At a point in the not distant future—perhaps with the help of this book—practically every human being on earth will know who and what they actually are, and here is something that is of utmost importance that will come about as a result: the world will become a much better place.

Let's Explode the Myth

Before who you are is revealed, it's important to address and explode the myth that continues to be perpetuated by science, the one that causes many to believe they are their physical bodies and their minds—what I prefer to call, "the mind-body complex." It is the basic tenet of science that's taught in our schools and colleges today that nothing exists except material substance—matter. This is the supreme falsehood that continues to be put forth in spite of the findings of quantum mechanics that matter, as Scientific Materialists think of it, actually does not exist.

If the tenet were true, then mind, intelligence, and consciousness could not have existed until evolution produced a brain. It would follow that you are comparable to a robot with a computer-like brain, that your mind-body complex is you, and that when you die, that would be the end of your existence. But that will not be the case because you are not your body, and you are not your mind. You are consciousness.

Let me repeat that. You are consciousness itself.

Why am I so sure? Your consciousness is the one thing from which you cannot stand apart and observe. Consciousness is you, the observer. To repeat, you are the observer, the one who can control the actions of your body and your mind—the decision-maker. Consciousness gives you the sense, "I exist—I am."

Knowing this will give you power because it puts you in control. Because you are not your mind, but rather, the controller of your mind, you can consciously observe your thoughts and sift through them. You can discard some and keep others. What you focus on will grow and expand, which means you can decide what to do with your life. You can be whatever you want to be.

You—consciousness—animate your body. Some would refer to you as the "life force," life itself, and that is another way of describing what you are. You will continue living—be the observer—when you leave your body. This is what millions who have been clinically dead and been brought back to life have learned and will gladly tell you. Your body some day will die, but you—the real you—will not. Once you have moved on to other realms of existence, your body will decay—as in, "ashes to ashes, dust to dust"—but you will continue to for eternity.

Consciousness, as we will see, is the ground of being that creates everything, and you are an integrated part of and at one with it. You are what might be thought of as a spark of the eternal flame. Without consciousness, you would not exist, and in fact, without consciousness, nothing would exist. The implications of this are enormous. For example, not only are you eternal, as alluded to above, the potential you possess for self-actualization is limitless. But before we get more deeply into that, it's important to understand why what has been written above is true, so let us begin.

Consciousness Is Primary

Two mysteries have captivated the human imagination for thousands of years. The first is why the universe exists at all. Why is there something rather than nothing? The second is that conscious beings exist to perceive it. An ancient idea is that the mystery of consciousness and the mystery of existence are intimately connected, and today, after a hiatus of two or more centuries, a growing number of philosophers and scientists are taking this possibility seriously.

Back in the nineteenth century when Scientific Materialism came to dominate scientific thinking, it was thought that the universe had always existed. The universe was also thought to be much smaller than we now know it to be. It is a fact that most of our knowledge about the universe in which we live was gained during the 20th century. With the help of photography, fainter objects were observed. The Sun was found to be part of a galaxy made up of more than ten billion stars.

The existence of other galaxies was discovered by Edwin Hubble, who identified the Andromeda nebula as a different galaxy, and many others at great distances. Using powerful telescopes, astronomers were able to see that there are many galaxies, and due to their shift toward the red end of the spectrum of light, that those farthest away are moving away from us faster than those in closer proximity. As a result, cosmologists are able to peer deeply into the past and infer the state of the universe in what is thought to be its first fractions of a second—what has come to be known as, "The Big Bang." But where did everything come from? What created it, and what existed before the beginning?

Physicists have proposed that the spark of existence had its origin in a quantum fluctuation, triggering an explosive chain reaction, leading to the still evolving universe we inhabit today. This narrative, however, presupposes the laws of quantum mechanics. As British Biochemist Rupert Sheldrake said in a now banned TED Talk, "[Scientists today say] give us one free miracle and we'll explain the rest.' And the one free miracle is the appearance of all the matter and energy of the universe, and all the laws that govern it, from nothing in a single instant."

Suffice it to say that rather than explaining existence, current scientific theories of the origins of the universe have simply pushed things back to a point that raises the question, "What existed before the beginning?" Could it all have come from nothing? Although that is what many scientists purport to believe, it doesn't make sense. As the song in *The Sound of Music* goes, "Nothing comes from nothing, nothing ever could."

Instead of beginning with nothing, it seems logical that the challenge of explaining existence should focus instead on defining a self-existing ground of being for which no explanation is required. Some physicists have proposed that the true ground floor of reality is the seething quantum realm of particles, forming in and out of existence. While this level of reality surely exists, there is no clear reason why the primordial situation should be constrained by quantum physics. A deeper level of explanation seems to be required, and what makes sense is that consciousness is the ground of being. How seething quantum particles came to be the ground of reality calls out for an explanation, but in theory, consciousness can explain itself. A unique feature of consciousness is that it does not appear grounded in anything beyond itself. The conscious self is self-producing in so far that it exists only in and to itself. As René Descartes [1596-1650] famously said, "I think therefore I am." In other words, nothing is required beyond consciousness for existence to be a demonstrated fact.

I intuit someone reading this saying to him or herself, "Wait a minute. The brain creates consciousness."

Not so. Although it hasn't been publicized for reasons stated in the first paragraph of this book—that scientists want to keep you in the dark order to save face and maintain power—the conclusion drawn after sixty years of exhaustive research at the University of Virginia School of Medicine is that the brain does not create consciousness. The brain is a receiver of consciousness like a cell phone of text and voice messages—a receiver that integrates consciousness with your body, which as mentioned previously, might better be called your "mind-body complex," a vehicle or apparatus that allows your consciousness to inhabit the physical dimension. If you don't believe this, go to YouTube and put the following phrase in the search bar: "Dr Bruce Greyson Consciousness Independent of the Brain." A lecture by Dr. Greyson that goes into detail about UVa's research should appear at or near the top.

Dr. Greyson, by the way, is not some New Age Looney Tune. He is Professor Emeritus of Psychiatry and Neurobehavioral Sciences at the University of Virginia. Additionally, in the Appendix of this book you will find Chapter One from my book, *Life After Death, Powerful Evidence You Will Never Die,* summarizing Dr. Greyson's lecture.

By the way, if the brain does not create consciousness, where do you suppose it comes from? Consciousness is the Source, the Creator, the ground of being that under-lies, supports and informs reality, and your personal por-tion of it was molded and formed during countless incarnations. But before we explore that, let's look at a few facts that support the contention that consciousness is primary.

Quantum Mechanics and Consciousness

While Scientific Materialists continue to maintain that only matter exists, Albert Einstein told us that what we normally think of as matter does not exist. He did so with the formula, $E=MC^2$, according to which energy equals mass times the speed of light, squared. In other words, mass (matter) and energy are the same in different forms. Moreover, it is a demonstrated fact that con-sciousness and matter interact. A quantum mechanics experiment known as "The Double Slit Experiment," for example, indicates that observation by a researcher, i.e., consciousness, changes waves of light into particles.

In 1803, Thomas Young (1773-1829) demonstrated that light is waves by placing a screen with two parallel slits between a source of light—sunlight coming through a hole in a screen—and a wall. Each slit could be covered with a piece of cloth. These slits were razor thin, not as wide as the wavelength of the light. When waves of any kind pass through an opening not as wide as they are, the waves diffract. This was the case with one slit open. A fuzzy circle of light appeared on the wall. When both slits were uncovered, however, alternating bands of light and darkness appeared, the center band being the brightest. Scientists call this a zebra pattern. The areas of light and dark result from what is known in wave mechanics as interference. Waves overlap and reinforce each other in some places and in others they cancel each other out. The bands of light on the wall indicated where one wave crest overlapped another crest. The dark areas showed where a crest and a trough met and canceled each other out.

In 1905, Albert Einstein published a paper that revealed light also behaves as if it consists of particles. He did so by using the photoelectric effect. When light hits the surface of a metal, it jars electrons loose from the atoms in the metal and sends them flying off as though struck by tiny billiard balls.

Now let's consider a double slit experiment constructed to determine what happens when those conducting the experiment observe or do not observe which slits the photons of light pass through. In this experiment, a gun is used that fires one photon at a time. Both slits are left open and a detector is used to determine which slit a photon passed through and where each it hit. The photons make marks, tiny dots, on a screen. Only one photon at a time is shot, so that there can be no interference.

When the detector is turned off, and it cannot be known which slit a photon passed through, something counterintuitive, you might even say "mind-blowing" happens. The zebra pattern appears. In other words, without the detector making it possible for the researcher to observe which slit a particle passed through and where it one hit, the particles behaved like waves even though they are fired one at a time. Rather, they behave as if they were all fired at once.

The bottom line of this experiment is that the researcher's ability to know—*consciousness*—causes the waves to collapse into particles that form a pattern. This suggests several things: 1) that consciousness is the ground of being and the medium of the single mind we all share. As a result, when the researcher—whose mind is integrated with the single mind—can access knowledge about the photons, the zebra pattern does not occur. 2)

Observation, i.e., consciousness, creates reality. Without consciousness only potential exists—in this case, in the form of light waves. When consciousness enters the equation, potential in the form of waves becomes particles. 3) Without observation, i.e., consciousness, the experiment takes place outside four-dimensional (height, width, depth, time) reality, a realm in which time does not exist. Without observation, i.e., without consciousness, it appears that all the photons were fired at once, rather than one at a time, thereby creating a zebra (wave interference) pattern.

These phenomena were examined and verified by setting up the experiment several ways, and they have been repeated and confirmed in a number of laboratories. In the first, the detectors were in front of the two slits. In the second, researchers placed detectors between the screen and the two slits, i.e., after the photons had passed through them. As in the original experiment, knowing about a photon's behavior at the two slits made the zebra pattern vanish, whether or not the detectors were before or after the slits (see the accompanying graphic). But when the detectors were switched off, the zebra stripes returned.

In a third variation, a detector was placed before the slits and a mechanism erased the knowledge after the photon had passed through. The same thing happened. The zebra pattern returned. The result was the same no

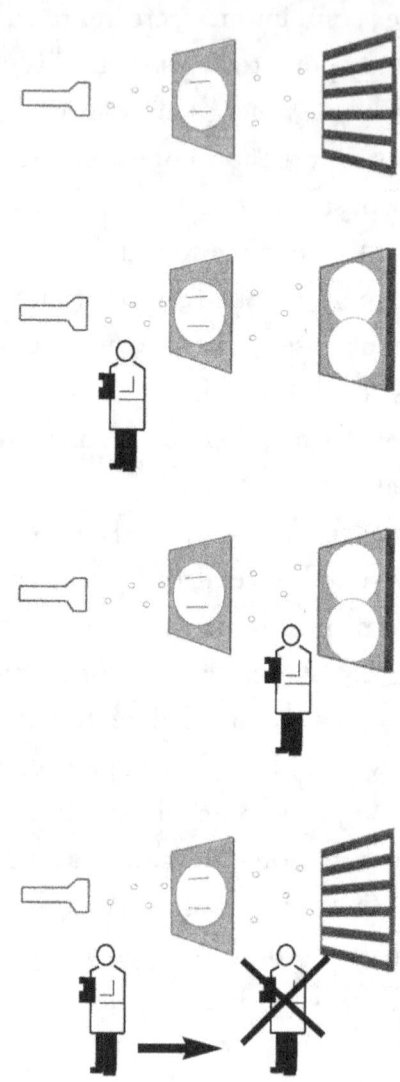

The Double Slit Experiment demonstrates the observer's mind is at one with the Infinite Mind.

matter which way the experiment was set up—before the slits, after the slits, or before the slits and then erased. Whether or not the researcher was able to know where each photon hit determined the presence of the zebra pattern, or the lack of it.

Matter from Mind

In the experiment discussed above, consciousness was shown to affect physical reality, but can it create physical reality?

Apparently it can, according to evidence that was collected by a college professor named Stephen Braude, whom I also interviewed for a weekly podcast I once hosted and produced. At the time, Dr. Braude was a tenured professor of philosophy at the University of Maryland Baltimore County, and I had just read his book, *The Gold Leaf Lady and Other Parapsychological Investigations* (The University of Chicago Press, 2007).

Dr. Braude related several well documented and amazing stories of mind over matter, but perhaps the most fantastic, as well as the one that supports the contention that mind [Infinite Mind] creates matter, had to do with Katie, a woman born in Tennessee, the tenth of twelve children.

Katie is apparently a simple woman. She was illiterate, at the time Dr. Braude wrote the book about her, lived in Florida with her husband and worked as a domestic. She was also a psychic who'd had documented successes helping the police solve crimes. In one instance she was able to describe the details of a case so thoroughly and accurately, the police regarded her as a suspect until those actually responsible were apprehended. She apparently also was able to apport objects—in other words, she somehow caused them to disappear in one place and reappear in another, at least that is what Dr. Braude maintained when I spoke with him. And that wasn't all. Seeds reportedly germinated rapidly in her cupped hands. Observers claim to have seen her bend metal, and she was both a healer and a medium or channel. Being illiterate, she could not read or write in her native English, but she has been video taped writing quatrains in medieval French similar both in style and content to the quatrains of Nostradamus.

I know some scientists are going to have a hard time believing what comes next because it goes against what they see as a fixed law of physics—that matter cannot be created nor destroyed—but most amazing, perhaps, is what appeared spontaneously on her skin—on her hands, face, arms, legs, and back—apparently out of thin air. It looked like gold leaf, a thin version of the wrapping on a Hersey's Kiss. Katie could not control when this happened, but Dr. Braude and other witnesses saw the foil materialize firsthand. He even videotaped it appearing on her skin.

I just stopped typing and checked. As of this writing, footage from this video can be seen on YouTube. Go to YouTube and put "gold leaf lady Braude" in the YouTube search bar. Several videos about this will come up. The title of the video with footage of Katie and her gold leaf is "UMBC In the Loop: Stephen Braude."

Dr. Braude took the foil to be analyzed. It turned out not to be gold at all, but brass—approximately 80 percent copper and 20 percent zinc.

Dr. Braude thinks there's a reason she produces brass and not gold. Where does the brass foil that appears on Katie's skin come from? It appears that her mind creates it. In fact, as mentioned, Dr. Braude believes she produces brass rather than gold for a reason. You see, Katie had a difficult and tense relationship with her husband. Once she apported a carving set. It just appeared. And her husband—apparently nonplussed—said, "So what? It's not worth anything." Soon afterward, gold colored foil began appearing on Katie's skin. But it wasn't real gold, it was fool's gold—brass. Dr. Braude thinks this is how she gets back at her husband. Katie's mind—albeit the unconscious part—creates matter in the form of brass foil. This being the case, why should it be difficult to believe that an Infinite Mind—one infinitely more powerful than a human mind—created the material universe? The physical universe had to come from something.

Chapter Two: The Facts of Life

If matter is all that is, as traditional scientists continue to claim, as mentioned previously, consciousness and intelligence could not have existed until evolution produced a brain. If that were true, then everything in physical reality, from life itself to the laws of physics, would have to have come about by accident.

Does anyone in the twenty-first century who has really thought about it still think that's the case? Although I doubt it, it is what many people came to believe after Darwin published The *Origin of Species* in 1859. By any measure, that was a long time ago, and was a time in history when science was still in what can be described as a primitive state because the knowledge back then of how things worked was far behind what it is today. For example, scientists did not yet know that bacteria cause disease. Bacteria were thought to be a symptom rather than a cause. Scientist back then also thought that atoms that form matter were comparable to tiny marbles—rather than packets of rapidly vibrating units of energy now known as electrons and quarks. Back then life was thought to have come about by accident when lightning struck a primeval lagoon made up of the right combination of chemicals. Amazingly—and I find this to be incredible—that's still what scientists today purport to think. Apparently, those who know the truth either keep quiet about it for fear of being ridiculed by their peers, or they do so to save face and maintain power, or because they think it might lead to a resurgence of religion, which they sincerely hope will go away or disappear.

A lot happened and was discovered between 1859 and 1957 that increased our knowledge—things like quantum mechanics, microbiology, and an expanding universe that appears to have had a beginning—but even so, let's fast-forward to that date. It's the year Francis Crick [1916-2004] discovered that the chemical subunits along the interior of the double helix of DNA function like alphabetic characters in a written language, or the digital characters such as the zeros and ones in a computer code. No doubt you've seen DNA code printouts. Crick realized they direct the construction of proteins and protein machines that all cells need to stay alive. In other words, it came to light that digital information directs the construction of the crucial components of living cells. Therefore, to explain the origin of life, one would have to explain how this complicated processing system came about.

How complicated is it? According to an article on the website of *BBC Science Focus Magazine*, the UK's leading science and technology monthly: "The DNA in your cells is packaged into 46 chromosomes in the nucleus. As well as being a naturally helical molecule, DNA is super-coiled using enzymes so that it takes up less space. If you stretched the DNA in one cell all the way out, it would be about two meters long and all the DNA in all your cells put together would be about twice the diameter of the Solar System."

How incredible is that? The strand of DNA in a single cell is six feet, six inches long. That must contain an awful lot of code when you consider that the size of the characters in the code is microscopic. Think of the enormous amount of information packed into it. How in the world could lightning striking the chemical soup in a primeval lagoon result in a microscopic strand of computer-like code more than six feet long?

The answer is, it couldn't.

So that you are able see for yourself that I'm not making this up, here is a link to the article just referenced above:

https://www.sciencefocus.com/the-human-body/how-long-is-your-dna/

It should go without saying that whenever we see information and trace it back to its source, whether it's computer code, a paragraph in a book, or a computer program, there is always an intelligent input that accounts for that information. This indicates, of course, is that intelligence is behind the origin of life, and yet an ardent Scientific Materialist would argue that given infinite time, anything can occur by accident—for example, that a room full of monkeys with typewriters would eventually produce *War and Peace* or the complete works of Shakespeare with no typos. It would only be a matter of time. The problem is that mathematicians who have worked out the odds of that actually happening say it would take something like infinity in terms of time and an almost infinite number of monkeys and typewriters—in other words, it is virtually, if not completely impossible.

The fact is, DNA could not have happen by chance. There hasn't been enough time. Scientist say the universe began with a Big Bang 13.8 billion years ago, and the earth is about 4.5 billion years old. They say life began on earth 3.77 billion years ago, and so that leaves 730 million years for chance to have typed out six feet, six inches of microscopic code—with no typos. According to mathematicians who have worked out the odds of that happening, it simply could not have come about by chance. Of course, it's true some scientists continue to argue against the theory that the universe had a beginning. They believe the universe has always existed and that it contracts and expands. But whether it had a beginning or has always existed and contracts and expands, the result would be the same—it got off to a (perhaps new, after an infinite number of previous) start(s) 13.8 billion years ago. This is indicated by a broad range of phenomena, including the abundance of light elements, the

cosmic microwave background (CMB), large-scale struc-
ture and Hubble's law, i.e., the farther away galaxies are,
the faster they are moving away from Earth.

It seems to me that anyone with an I.Q. above 65 who
thinks deeply about what is written above would have to
come to the conclusion that information resembling
computer code that directs something complicated to
happen must be the product of some sort of intelligence.
And yet that would have been impossible if material substance—matter—is all there is. To repeat what has been
stated twice already, if that were the case, intelligence
could not have existed until evolution produced a brain.

Not every scientist has turned a blind eye to the
facts. After Click's discovery, a number of them began to
see that there must be some sort of guiding intelligence
responsible for the origin of life. I know this from personal experience because I read a book more than forty
years ago that put forth that argument. Published in 1975,
it refuted the idea that intelligence, consciousness, and
awareness, came about as a result of evolution. The book
was entitled *Intelligence Came First*. It was produced by a
group of well qualified individuals that met monthly to
read and discuss material that was compiled and edited
by Ernest Lester Smith [1904-1992], a Fellow of the
Royal Society—the prestigious scientific academy of the
United Kingdom, dedicated to promoting excellence in
science.

Intelligence Came First caused quite a bit of controversy when it came out. Among other things, the book put forth the DNA, computer-code argument above, and it also noted that throughout the eons of evolution, needs have preceded the organs through which they are fulfilled—eyes, ears, taste buds, hearts, kidneys, and so forth. Each new organ developed in response to a need, the book's contributors argued, so why would the brain be an exception? The conclusion the book's authors and contributors came to was that intelligence came first, quite able to function in its own realm.

What is the source of consciousness and intelligence that created all that is? The Source is the Infinite Mind of which each and all of us are part. It follows that you are a portion of divinity—if you would like to call it that—inhabiting a mind-body complex. This being the case, all knowledge is within you. All power is within you. This is something gurus and mystic have realized down through the ages. It is what Jesus knew and what gave him the power to work miracles.

Let me say here, parenthetically, to any scientist or atheist reading this, that I am not writing this book to promote religion. I am writing it to get the truth out to anyone who may be interested. I happen to think, however, that Jesus knew what he was talking about, just as many mystics down through the ages—including the Buddha—have known what they were talking about. Unfortunately, what Jesus said was largely misunderstood when he said it 2000 years ago, and it continues to be misunderstood by many today.

Christians believe that Jesus was God incarnate, and that is true. But it is also true that each and every one of us—you included—is God incarnate, assuming you want to label the Infinite Mind, "God." Consider this. In Chapter Ten of the Gospel of John, Jesus explained that it wasn't he but the "Father," i.e., what Christians typically assume was God working through him that caused the miracles with which he was credited. As part of his explanation he said, "I and the Father are one." (See John 10:30 NIV.) This got him into trouble with Jews who became angry and were about to stone him.

Jesus replied to the angry mob by saying, "I have shown you many good works from the Father. For which of these do you stone me?" (John 10:32 NIV)

The Jews answered, "We are not stoning you for any good work, but for blasphemy, because you, who are a man, declare yourself to be God." (John 10:33 NIV)

Jesus then quoted Psalm 82:6: "Is it not written in your Law: 'I have said you are gods'?" (John 10:34 NIV)

By quoting this Scripture Jesus clearly was indicating that the Jews who wanted to stone him were "gods," as was he and every other human being—and that includes you regardless of the color of your skin or whether you are a Christian, an atheist, or a follower of Islam. The reason people think it is impossible that God is within us and that we, like Jesus, are God, is that for millennia our culture has bought into the idea that God can be compared to a man with a long white beard lounging on a cloud up in the sky, i.e., that he is a separate entity. But he is not. He is everywhere—the ground of being. All truly is one, seamless whole. It is impossible to go anywhere where God—Infinite Mind—is not.

I invite Christians who may scoff at this, or who think I am the one guilty of blasphemy, to consider these words also spoken by Jesus: "Whoever believes in me will do the works I have been doing, and they will do even greater things than these . . . " (See John 14:12 NIV) To believe in Jesus means, among other things, to believe what he said and taught is true, and that requires believing that we must have the ability, as he said in the Scripture just quoted, to perform miracles. Once we fully become fully evolved beings, as Jesus was, we will be able to do so.

Jesus also famously said, "Anyone who has seen me has seen the Father" (See John 14:9). Christians take that statement to mean Jesus was saying he was God incarnate. And he also said, "When you have done it unto the least of these you have done it unto me." (See Matthew 25:40.) Christians have a difficult time figuring out what he meant by that, but it becomes obvious when those two statements are put together that Jesus understood that we all are one, and that the "I AM" at the back of our minds is the window on the One Life that is each and every one of us.

Jesus was all, just as I hope you will come to recognize that you are all.

How can you wrap your mind around this? How do you come to recognize and feel at a deep level that there is no separation from you, others, and God? I suggest you spend time every day when you're out in the world seeing other people as simply yourself in another body. That may seem difficult and perhaps silly at first, but after a while, it will do wonders for opening your heart to the world and to universal love. Seeing others as an extension of yourself will bring peace of mind and help mitigate your suffering in difficult situations. Moreover, you cannot truly serve others with deep compassion unless you see them as part of yourself.

Let me interject here that once you realize you and others are one, you do not have to become a pushover or a doormat. If people want to steal from you, or loot your business, for example, it is correct to stop them. Oneness means that others are no less the Creator than you, but it also means they are no more the Creator than you. Treating others as you want to be treated is the right way for everyone to behave, and that means that if someone treats you in a way you don't want to be treated, you are obligated to speak your truth and stand up for yourself.

All Is One

What I am hoping to communicate is that All-Is-One, and your consciousness, my consciousness, and indeed everyone's conscious is the single, unified, underlying "I AM" consciousness that Jesus called his "Father." It is the infinite consciousness that underlies, supports, informs—and indeed creates physical reality. You don't have to be religious to understand that—you just have to open your eyes and your mind. It is what Jesus understood, and that understanding along with powerful belief is what gave him the ability to work miracles. When that has been taught in school for a couple of generations, and everyone realizes it, racism, poverty, and suffering will become greatly diminished.

Perhaps you are thinking, "All-Is-One? Really? Come on, I don't buy that."

If that's you, read Gary Zukav's book, *Dancing Wu Li Masters: An Overview of the New Physics.* In it, he explained quantum mechanics without using complicated mathematics. Consider, for example, the following statement from that book:

> *. . . the philosophical implication of quantum mechanics is that all of the things in our universe (including us) that appear to exist independently are actually parts of one all-encompassing organic pattern, and that no parts*

of that pattern are ever really separate from it or from each other.

Alan Watts [1915-1973], a twentieth century philosopher and interpreter of Zen Buddhism, answered children's questions concerning why they were here, where the universe came from, where people go when they die and so forth with a parable about God playing hide and seek. Watts told them God enjoys the game, but has no one outside himself to play with since he is All-That-Is. God overcomes the problem of not having any playmates by pretending he is not himself. Instead he pretends that he is me and you and all the other people and the animals and rocks and stars and planets and plants and in doing so has wonderful and wondrous adventures. These adventures are like dreams because when he awakes, they disappear. Here is some of what Watts wrote:

> *Now when God plays hide and pretends that he is you and I, he does it so well that it takes him a long time to remember where and how he hid himself. But that's the whole fun of it—just what he wanted to do. He doesn't want to find himself too quickly, for that would spoil the game. That is why it is so difficult for you and me to find out that we are God in disguise, pretending not to be himself. But when the game has gone on long enough,*

all of us will wake up, stop pretending, and remember that we are all one single Self—the God who is all that there is and who lives forever and ever.

It will no doubt be shocking to some to think of themselves as God, but Watts was talking about the core essence that is beyond the ego and deeper within than the personal unconscious, the collective unconscious, the archetypes and so on. As Joseph Campbell [1904-1987] said in the PBS TV series, *The Power of Myth,* "You see, there are two ways of thinking 'I am God.' If you think, 'I here, in my physical presence and in my temporal character, am God,' then you are mad and have short-circuited the experience. You are God, not in your ego, but in your deepest being, where you are at one with the non dual transcendent."

Chapter Three: Why the Truth about You Is Taboo

Why do the leaders of science continue to hold on to the Scientific Materialist model? Because they don't want to admit they have been wrong. Imagine you are a professor in the sciences at a university, or the editor of a science journal, someone who has written books and taught a couple of generations of students. Are you going to admit you have been preaching something that is patently incorrect?

Not if you can help it.

Why do some politicians not want the truth to come out? Because they stay in power by pitting one group of humans against another and playing the race card. If it came out that we are each a spark of consciousness, an unit of the divine, that all of us are united by a single mind that supports and informs everyone and every- thing, it would soon become clear to everyone that su- perficial differences such as the pigmentation of someone's skill are meaningless.

We indeed are all created equal.

What about religious leaders? Why don't they want the truth to come out? Wouldn't it help their cause?

No, it would not for two reasons: 1) The Abrahamic religions (Christianity, Judaism, and Islam) assume God and his creation are two different things when the reality is that All-Is-One. 2) Reincarnation was purged from

Christianity at the Second Council of Constantinople in 551, and it is not a tenet of Judaism or Islam. On the other hand, eastern religions such as Hinduism and Buddhism embrace reincarnation, and the reality of reincarnation is one of the products of the understanding that humans are manifestations of consciousness.

Is Reincarnation Really Possible?

Researchers at the University of Virginia have been conducting investigations into the recollection by children of past lives for about sixty years. As a result, they have in excess of 2500 cases in their files. I was quite familiar with this even before I saw the lecture by Dr. Greyson by mentioned earlier because of research I had done for my book, *REINCARNATION: Good News for Open Minded Christians and Other Truth Seekers.* I have in fact twice interviewed one of the Perceptual Division's key researchers who has written two books on the Division's reincarnation research findings, Jim B. Tucker, M.D., a Phi Beta Kappa graduate of the University of North Carolina, a medical doctor, and a board certified child psychiatrist who at the time I spoke with him served as medical director of the Child & Family Psychiatry Clinic at the University of Virginia School of Medicine.

Sixty Years of Irrefutable Research

Suppose you were changing your son's diaper—let's say he was just beginning to talk and was quite verbally adept at the age of 18 months—and he looked you in the eye and said, "When I was your age, I used to change your diaper."

What would you think?

If your father happened to be deceased, would you possibly think your son might be your father reincarnated? That would make him his own grandfather.

Can something like that happen? Dr. Tucker told me that it can.

At the time I spoke with Dr. Tucker, about 1600 of the 2500 cases—many of which came from Dr. Tucker's predecessor, the late Ian Stevenson, M.D. (1918-2007—had been entered into a computer database along with the information collected on each. The data were sorted into about 200 different variables, allowing researchers to comb through and cross tabulate the data to spot trends as well as to categorize and compare the similarities and differences based on various factors and characteristics.

Dr. Stevenson was a methodical and meticulous researcher who graduated first in his medical school class at Canada's McGill University. He never actually claimed reincarnation as fact, but rather, said his cases were "suggestive" of reincarnation. His often-cited first book on the subject was published in 1966 and entitled, *Twenty Cases Suggestive of Reincarnation.*

The cases he studied come from all over the world. When Dr. Stevenson began this research, they were easiest to find in places where people have a belief in reincarnation such as India and Thailand. This may be because parents were not as likely to think a child was imagining a past life, and because they are not likely to be embarrassed to talk about it. Nowadays, however, people in the United States are not as reticent as they once were. Dr. Tucker said that since the University of Virginia set up a web site on this subject some time ago, he and his colleagues hear from parents "all the time" about their children's memories of past lives.

Nevertheless, in the United States reincarnation is thought by many to go against Christian doctrine, even though recent surveys show that more than twenty percent of Christians believe in reincarnation. The percentage is higher, by the way, among younger adults.

Reincarnation and Christianity

I'd like Christians who may be reading this to know about a man I interviewed on my radio show in the spring of 2008. His name is James A. Reid Sr., and he's a Southern Baptist minister, now retired. He holds a Doctor of Ministry degree from San Francisco Theological Seminary. For 15 years he was Chaplain to the Los Vegas strip, where he heard a lot of talk about Edgar Cayce and past lives, which he always dismissed as fantasy thinking.

Finally, he got so fed up he decided to write a book denouncing reincarnation as a Biblically untenable doctrine. But Dr. Reid is an honest and mature individual. Once he dug into Church history and the Scriptures, he was forced to change his view. He ended up writing a book that maintains the Bible supports the doctrine of reincarnation. It is called, *BORN AGAIN AND AGAIN AND AGAIN: A Bible-Based View of Reincarnation.*

Dr. Reid maintains that for the first five hundred year history of the Church, many accepted reincarnation as fact. It wasn't until 553 A.D. that it was condemned by the Second Council of Constantinople, and then only by a narrow margin. He gives several examples indicating Jesus and others of his time believed in reincarnation. For example, John the Baptist was supposed by many to be the prophet Elijah reincarnated. Jesus himself said this was so. (See Matthew 11:14.) Once, Jesus asked his followers who people thought he (Jesus) was. They replied that many believed him (Jesus) to be one of the prophets—presumably reincarnated, since the last prophet died about 400 years earlier. Also, consider the story of Jesus restoring the sight of the man who had been born blind, as recounted in John 9:1-12, in which Jesus' disciples ask him if the man's sins caused his blindness, or if the sins of his parents had caused him to be born blind.

Since the man was blind from birth, the only way his own sins could have caused his blindness was for him to have sinned in a former life. Jesus did not tell his followers this wasn't possible. To the contrary, he seems to have assumed it was possible, although he gives another reason for the man's blindness, saying, "Neither this man nor his parents sinned, but this happened so that the work of God might be displayed in his life."

Edgar Cayce, whose psychic readings probably did more than anything to promote the concept of reincarnation in the West, was a devout Presbyterian and Sunday school teacher who read the Bible once through for every year of his life. At first, when reincarnation started showing up in his readings, he was baffled and confused. But he reread the Bible and satisfied himself it wasn't anti-Christian.

There are many references to reincarnation in the Bible but believers overlook or misinterpret them because they have been conditioned to think reincarnation is taboo. Kevin Todeschi, Executive Director of the Association for Research and Enlightenment, said on my radio show in November 2007, that he has counted eleven such references in Matthew's gospel alone.

The Child Who Is His Grandfather

In the case mentioned at the beginning of this chapter—the 18-month-old child who said he had changed his father's diaper when he was his father's age—the child's mother was the daughter of a Southern Baptist preacher. As you might imagine, she found what her son said to be highly unusual. I asked Dr. Tucker to describe the case when he came on my show, and he obliged.

The child's grandfather had died eighteen months before the child's birth. His first mention of having been his own grandfather was during that change of diapers, but as time went by he made more comments about how he used to be big, and what he did when he was. His mother in particular became interested and began to ask the boy, whose name was Sam, questions. Sam came up with some very specific statements. For instance, she asked him if he had had any brothers or sisters. He said he had had a sister who was killed. In fact the grandfather's sister had been murdered sixty years before.

The parents felt certain the child could not have known this since they had only recently learned about it themselves.

The child also talked about how, at the end of his previous life, his wife would make milkshakes for him every day, and that she made them in a food processor rather than in a blender. This turned out to be true.

When Sam was four years old, his grandmother—his wife in his previous life—died. Sam's dad traveled to where she lived and took care of the estate. When he returned, he brought some family photos with him.

One night Sam's mother had the pictures spread out on the coffee table. Sam walked over and pointed to pictures of his grandfather and said, "Hey, that's me."

To test him she pulled out a class photo from the time the grandfather was in elementary school. Sam ran his finger across the photo, which had sixteen boys in it, and stopped on the one who had indeed been his grandfather.

"That's me," he said.

The Reason Sam May Have Come Back

The grandfather may have come back as the son of his own son because of the relationship—or lack thereof—the two had had in his previous life. The grandfather had not had an open relation with Sam's dad. He had been a very private person. Sam's dad felt that if his father had really returned as his son, his father may have decided to come back to try to develop a closer bond than had existed in their previous relationship. Dr. Tucker said this may be true. When he visited the family he could see that Sam and his dad were very close.

A Murder Victim Comes Back

Another story from the files of the University of Virginia, Dr. Tucker related on my show has to do with an Indian girl named Kum Kum, who said she had been murdered in her previous life—poisoned—by her daughter-in-law. Kum Kum said she was from a city of about

200,000 located about 25 miles away. One of the things that makes this a good case is that her aunt wrote down a number of statements—eighteen in all—she made before an effort was undertaken to see if they checked out.

All of them did.

The statements included the name of a son, the name of a grandson, the fact that the son had worked with a hammer. And a number of other specifics—for example, that she had a sword hanging near the cot where she slept, and a pet snake she fed milk to.

Research led to the woman Kum Kum claimed to have been, who had died five years before she was born. A big family flap had taken place over a will and who would inherit the worldly possessions of the deceased woman's son. Kum Kum had probably been right. Circumstantial evidence indicated the son's wife had poisoned her mother-in-law—the woman Kum Kum insisted she'd been.

What Many Cases Have in Common

These case histories are fascinating and convincing, and we could go on almost indefinitely considering them, individually. After all, there are more than 2500 in UVA's files. Instead, let's step back and look at the overall findings of this exhaustive study.

Children who report past-life memories typically begin talking about a previous life when they are between two and three years old. Those who have studied this phenomenon believe that emotional involvement with past-life family members argues for reincarnation as the cause rather than superpsi at work—the psychic reservoir of memories housed in the Infinite Mind also known as the Akashic Records. The children tend to show strong emotional involvement with such memories and often tearfully ask to be taken to the previous family. Once there, not only is a deceased individual usually identified whose life matched the details given, during the visits, children often recognize family members or friends from that individual's life. Tearful reunions are common.

Birthmarks and Birth Defects Provide Evidence

Many children studied also had birthmarks that matched wounds on the body of the deceased individual. To give one example, a boy in Thailand, who said he'd been a schoolteacher in this previous life, was shot and killed when riding his bicycle to school one day. He gave specific details including his name in that life and where he had lived. He continued to make this claim until his grandmother took him to the previous address. The child was able to identify the various members of his previous family by name.

Even more startling, he was born with two birth marks: a small round birthmark on the back of his head and a larger, more irregularly shaped one near the front.

The woman he claimed was his wife in that life recalled investigators saying her husband had been shot from behind. The investigators said they knew this because he had a typical, small, round entrance wound in the back of his head and a larger, irregular exit wound in front.

In another case, a boy remembered a life in a village not far away in which he had lost the fingers of his right hand in a fodder-chopping machine. The child was born with an intact left hand but the fingers of his right hand were missing.

The average length of time between the death and rebirth of the children in these birthmark cases is only fifteen to sixteen months. Somehow, the "veil of forgetting" most of us experience when we incarnate did not take hold. Perhaps, some children retain memories of the most recent past life because their souls have taken a shortcut between lives, skipping a process by which the life just lived would have been fully integrated into the soul.

Twenty-Two Percent of UVA Cases Have Birth Defects

According to Dr. Tucker's book, *Life Before Life* (St. Martin's Griffin, 2005), about 22 percent of the cases in the University's database include birth defects due to wounds suffered in violent deaths in the previous life. Most of the cases come from the Hindu and Buddhist countries of South Asia, the Shiite peoples of Lebanon and Turkey, the tribes of West Africa, and the tribes of northwestern North America.

In 1997 Stevenson published details of 225 cases in a massive work *Reincarnation and Biology: A Contribution to the Etiology of Birthmarks and Birth Defects.* The same year he presented a summary of 112 cases in a much shorter book, *Where Reincarnation and Biology Intersect.*

In many cases postmortem reports, hospital records, or other documents were located and consulted that confirmed the location of the wounds on the deceased person in question matched the birthmarks. These often correspond to bullet wounds or stab wounds, and as in the case described above. Sometimes two marks correspond to the points where a bullet entered and then exited the body.

Birthmarks also related to a variety of other wounds or marks, not necessarily connected with the previous personality's death, including surgical incisions and blood left on the body when it was cremated. A woman run over by a train that sliced her right leg in two was reborn with her right leg absent from just below the knee. A man born with a severely malformed ear had been resting in a field at twilight, mistaken for a rabbit, and shot in the ear.

Behavior Traits Also Provide Evidence

Further evidence for reincarnation comes from what might be called behavioral memories. For example, cases exist where children of lower caste Indian families believe they had been upper class Brahmins, and in their view still were. These children would refuse to eat their family's food, which they considered polluted. Conversely, a child remembering the life of a street-sweeper—a very low caste—showed an alarming lack of concern about cleanliness. Some children demonstrate skills they have not learned in their present life, but which the previous personality was known to have had. A number of Burmese children who recalled being Japanese soldiers killed there during World War Two preferred Japanese food such as raw or semi-raw fish over the spicy Burmese fair served by their families.

Many children express memories of the previous life in the games they play. A girl who remembered a previous life as a schoolteacher would assemble her playmates as pupils and instruct them with an imaginary blackboard. A child who remembered the life of a garage mechanic would spend hours under a family sofa "repairing" the car he pretended it to be. One child who remembered a life in which he had committed suicide by hanging himself had the habit of walking around with a piece of rope tied round his neck.

Phobias May Originate in a Former Life

Phobias occur in about a third of the cases and are nearly always related to the mode of death in the previous life. For example, death by drowning may lead to fear of being immersed in water; death from snakebite may lead to a phobia of snakes; a child who remembers a life that ended when he was shot may display a phobia of guns and loud noises. A person who died in a traffic accident may have a phobia of cars, buses, or trucks.

Sexual orientation may also be affected by a previous life. In one of his books, Ian Stevenson wrote, "Such children almost invariably show traits of the sex of the claimed in the previous life. They cross-dress, play the games of the opposite sex, and may otherwise show attitudes characteristic of that sex. As with the phobias, the attachment to the sex and habits of the previous life usually becomes attenuated as the child grows older; but a few of these children remain intransigently fixed to the sex of the previous life, and one has become homosexual."

Certain preferences and cravings can also carry over. They frequently take the form of a desire or demand for particular foods not eaten in the child's present family, or for clothes different from those ordinarily worn by the family members. Other examples include cravings for addictive substances, such as tobacco, alcohol, and other drugs that the previous personality was known to have used.

UVA's Cases May Not Be Representative of the Whole

Dr. Tucker pointed out in one of my interviews with him that the cases he and others have studied may not be typical because most children do not remember past lives. As mentioned, the average time between lives in these cases is only fifteen months or so—although there are outliers that range up to fifty years. In 70 percent of these cases, the previous personality died by unnatural means, and many died young. This may speed up the reincarnation process. The consciousness may come back quickly due to unfinished business, or because he or she feels shortchanged. The quick return may also be the reason past life memories are intact, as well as sexual preferences, cravings and so forth. My guess is that a much longer duration between lives is the norm. Teachings of the Rosicrucians, a mystical order in which I have achieved the rank of Adept, say the human personality span under normal circumstances is about 140 years. If we live 70 years, for example, we can expect to spend 70 years in the realm between lives before we incarnate again. If we live 60 years, we can expect to spend 80 years between lives. The teachings stress, however, that this is a rule of thumb. Centuries could elapse between incarnations, or as with many in the UVA study, the return could come in a matter of months.

Children Aren't the Only Ones Who Have Past Life Memories

Memories of past lives also sometimes occur in adults, and such memories can be of lives that took place long ago. I once recalled a romantic interlude from a life as a Russian army officer during the Napoleonic Wars. The memory was triggered when I met someone that appeared to be same woman. A guest on my show recalled having a spontaneous recollection of a life as a woman that took place in twelfth or thirteenth century France. He said the woman he had been was being burned at the stake. He said this was so vivid it seemed more than a memory. He actually felt he was there, subjectively experiencing the ordeal.

He'd been meditating when suddenly it seemed he was back in the skin he'd occupied then—the action taking place around him. Information about who he was and what was taking place was present in his mind as though he had literally been transported back in time and reentered that body. He said that, surprisingly, he did not feel much pain at being burned—his consciousness exited his body as soon as the flames engulfed it. He floated nearby observing his body burn—not feeling any pain at all. Nevertheless, it was a gut wrenching, emotional experience that left him so distraught he secluded himself after reliving the experience and was unable to communicate with others for two or three days.

He said his need to withdraw had not been because of the pain he'd endured. It was a result of the distress he felt over the pain and suffering humanity puts itself through—man's inhumanity to man. He'd been burned at the stake in that life because he'd been a priestess of the Cathar religion, a Gnostic Christian sect persecuted and eventually extinguished by the Roman Catholic Church. His death by fire was just one of many that took place during the twelfth through fourteenth centuries.

Why were these people killed? No doubt they were seen as a threat to those in power, which at that time included leaders of the Church. The Church taught and teaches that salvation comes through belief in Christ and his sacrifice on the cross. Gnostics followed Christ's teachings but believed salvation comes through direct knowledge of God—a direct and personal relationship. Today, many if not all churches foster this direct relationship—a daily walk with God is considered a requisite by most. Catholicism in the time of the Cathars, however, taught that only the clergy could have this direct relationship.

The Past Lives of Glenn Ford

Anyone with an open mind who looks into what has been found will find it difficult to refute that reincarnation can and does happen. There are many anecdotal accounts. If you are old enough, you may remember Glenn Ford (1916-2006), a movie actor who flourished during

Hollywood's Golden Age. Once, when he was approached about taking a role in a movie about Dutch psychic Peter Hurkos (1911-1988), Ford decided he ought to learn something about the paranormal. So, at the age of 54, Ford personally viewed demonstrations by Hurkos, conducted a number of interviews with experts on the subject, and in December 1975, voluntarily underwent three past-life regression sessions via hypnosis, during which he described five previous lives. As is typically done in such sessions, the hypnotized subject, Glenn Ford in this case, was regressed back to childhood, and then was coaxed further back before his current birth to recall previous lives.

In one session, Ford described himself as a bachelor music teacher named Charles Stewart of Elgin, Scotland who died in 1892. Stewart loved horses but hated his job teaching music to young schoolgirls. Amazingly, under hypnosis, Ford agreed to demonstrate his musical skill, and played passages from Beethoven, Mozart, and Bach. When Ford listened to the tapes of the interview, he said he shared Stewart's love for horses and had, since his early years, been considered a natural with those animals. Most significantly, however, he said that in his current life he did not, and could not, play the piano. Following the past-life regression sessions, researchers went to Scotland and located historical records of a music teacher named Charles Stewart of Elgin, Scotland who died in 1892.

A second hypnotic regression session with Ford brought out memories of a life as a member of French King Louis XIV's elite horse cavalry. Under hypnosis, Ford not only gave accurate information about his surroundings in France 300 or so years prior, he was able to speak French fluently—although in his current incarnation, he did not know the language.

Subsequently, Ford was regressed to other previous lives, describing a Christian martyr killed by lions in the Coliseum in third century Rome, and a seventeenth Century Royal Navy sailor who died of the Great Plague. In his most recent prior lifetime, Ford was a cowboy who herded cattle in the American West. It is interesting to note that although Ford starred in 106 movies, as well as several TV series ranging from comedy to police dramas to war stories, he was best known for and most often cast as a cowboy in Westerns.

What does this suggest about you? You began as a spark of the consciousness of the Creator and have been evolving ever since. You likely first incarnated as a microbe and again in every stage between that and when primates evolved. In the millions of years since the early ancestors of modern humans first walked planet earth, you have probably lived dozens of times, perhaps hundreds, and maybe even thousands. If that is indeed the case, and it seems likely to me, you have been evolving, sometimes rapidly, and in some lifetimes, perhaps not so much. You have no doubt played many roles, from loser to winner and from warrior to wise elder. Everything you have experienced and learned is recorded in your subconscious mind—what Christians would call your soul.

Here's something else to realize and internalize: In some lives you have probably had black skin, in others white, and in some it may have been brown. Anthropologists and paleontologists tell us humans first evolved in Africa and that much of human evolution occurred on that continent. You have likely been around and evolving for eons, and so it seems unlikely if not impossible that your skin has always been the shade or color it is today. The fossils of early humans who lived between six and two million years ago, for example, come entirely from Africa. Most scientists currently recognize some 15 to 20 different species of early humans. This by itself should be sufficient to say that racism and tribalism make no sense at all. We are indeed all one.

Moreover, in some lives you have been a man and in others a woman. If you don't feel comfortable with the sex you came into this life with, perhaps in your most previous past life, you were a member of the opposite sex. If you have unexplained phobias, fears, or predilections—even talents as Glenn Ford had with horses—they very likely stem from events or conditioning that took place in a previous life. Research by UVa indicates this sort of thing is commonplace. Whatever the case may be, everything you have experienced is now part of you in what some would say is your subconscious mind, and what others would call your Soul.

Chapter Four: How You Create Your Reality

Having read to this point in this book, you now know that you are consciousness—the observer of your mind that has the power to pick and choose what you spend time and energy thinking about. This means you are no longer at the mercy of your thoughts, and you are no longer at the mercy of your upbringing. You are no longer at the mercy of anything that may have happened to you in this life, or in any other life you may have lived. As a result, you have the power to decide who you want to be, and what you want to do with your life.

No matter your background, race, or circumstances, the real you is perfect, and you must come to realize that, learn to love yourself, and understand that you do not need anything beyond yourself to be content. The truth is nothing actually exists beyond yourself. To think otherwise is to be caught in the illusion of separation created by the ego, which wants you to see the world and what is happening in terms of how it fits or does not fit the ego's emotional needs. In eastern thought this trick of the senses and the ego is seen for what it is: an illusion known as, "Maya."

To state the obvious, an ego is something you have that a little child does not. You have one because it developed and formed as you grew into an adult. A child views the world from the viewpoint of the "silent observer," the spark of divinity we have been discussing that connects you, me, and each one of us to the source. You viewed the world from this vantage point when you popped out of your mother's womb and opened your eyes for the first time. You continued in that state of being for quite some time, until eventually it faded away because your mind and your ego started getting in the way. William Wordsworth [1770-1850] wrote about this in his famous poem, *Ode: Intimations of Immortality from Recollections of Early Childhood.* Here is the first stanza:

There was a time when meadow, grove, and stream,
The earth, and every common sight,
To me did seem
Apparelled in celestial light,
The glory and the freshness of a dream.
It is not now as it hath been of yore;—
Turn wheresoe'er I may,
By night or day,
The things which I have seen I now can see no more.

Wordsworth was lamenting that he no longer saw the world through the eyes of a child, which we now know means he stopped seeing from the viewpoint of his true self, the silent observer. Because of that he lost the sense of peace and wonder he once felt. He did not know why, but we do. To repeat what I have been hoping to drive home, his ego and his mind got in the way.

All this is not to say the ego is a bad thing that should never have been allowed to happen in the first place. The development of ego was a necessary step in the evolution of the human species because our egos are what cause us to feel autonomous, and this is what gives us free will. The emergence of the ego and free will is what the allegory of Adam and Eve is about. By eating the fruit of the tree of the knowledge of good and evil, Adam and Eve became fully-fledged human beings. It gave them the ability to think objectively, to figuratively step outside themselves and observe their own behavior. It brought them into the four-dimensional world of duality. Before the snake, aka Eve's emerging ego, talked her into taking a bite of the apple, they were like every other creature in the Garden, forest, or jungle—non-dual creatures guided solely by instinct that walked and communed with the source in the cool of the evening.

It seems likely the source wants us to have free will because he/she/it experiences himself through us, and free will allows us to choose to do all sorts of things, good and bad, which in turn allows the source to experience a wide variety of adventures. This in turn allows him to come to know himself because the source is good and the source is love. One cannot know good without bad or love without fear or hate, just as one cannot know up without down or white without black. Moreover, I suspect the source wants us eventually to choose to be with and experience him/her/it without having been coerced, just as you would want your future spouse to have the freedom to choose you over others—because you are the one he or she truly wants. But for all the good the ego does, it also causes psychological suffering because it is the part of the mind that worries. It's what keeps you awake at night.

Worry was an important survival mechanism in prehistoric times. If a human didn't worry about where the next meal was coming from, or how to keep from freezing to death, or how to avoid getting eaten by a lion, he or she would probably not live long enough to reproduce and propagate the species. Suffering, it has been said, is "the catalyst of evolution" because we learn and evolve by facing and dealing with difficulties. Without hardship, there would be little incentive to advance and overcome difficult situations, and without suffering and sorrow, we would not develop compassion for others.

As noted above, our egos also create the illusion that we are separate from everyone and everything, and this can cause us to become barbaric and hostile toward others. But as we evolve, at some point we eventually learn the lessons of love and understanding and begin to remember that we are actually eternal beings having a temporary physical experience. That's when we start to realize that egotistical thinking can be counterproductive, and that "doing unto others as we would have others do unto us" is the best and most productive way to live.

Because you were attracted to this book and are reading it, I suspect you have reached, or will soon reach that state, which means it is time for you to transcend your ego mind and see it for what it is: a worry machine that has outlived its usefulness. It is time to return to the nondual world, which is the true nature of reality. This is what Jesus was talking about when he said in Matthew 6:25-27 (NIV), "Therefore I tell you, do not worry about your life, what you will eat or drink; or about your body, what you will wear. Is not life more than food, and the body more than clothes? Look at the birds of the air; they do not sow or reap or store away in barns, and yet your heavenly Father feeds them. Are you not much more valuable than they? Can any one of you by worrying add a single hour to your life?"

In order to stop paying so much attention to your ego, it's important to understand at a deep level that the ego can be compared to a computer program that has been built up over the course of your life. It must be clear to you that survival is its job, which is why it constantly causes you to waste time worrying.

The ego also causes you to react in often-predictable ways to situations that arise. For example, suppose someone says something to insult you? Let's say you bump into an old fraternity brother at the grocery store you haven't seen for a while—I'll call him Henry. And suppose Henry says, "Gee, Charlie, where did your hair go? Man oh man. Looks like you lost it somewhere."

Your ego will try to get you to react, and if you do, you might say, "I may have lost my hair, Henry, but I see you gained something—a huge beer gut. Or maybe not. Maybe you're really nine months pregnant. Which is it, Henry, you disgusting cur?"

The fact of the matter is that you don't have to react, and this is something very important to realize: A moment exists between stimulus and response when the real you can decide what to do. For example, why not just put aside the impulse to utter a comeback, and say, "Hi, Henry. Long time no see. Hey, there's a bar across the street. Let's go have a beer and catch up."

That you have the power not to be controlled by your ego is one reason knowing you are at one with the all brings a magnificent feeling. Alan Watts, who apparently had fully achieved this, wrote that a gut-level realization of oneness brings moments of joy that are incredibly intense. He also said that sorrows are looked upon philosophically and that the sense of union with the universe is empowering. Once you come into harmony with all that is, you will have arrived in position to achieve the totally fulfilling life you were born to live, and through effort and discipline, you will be able to achieve self-actualization.

Here is a direct quote from what Alan Watts wrote about this:

In immediate contrast to the old feeling, there is indeed a certain passivity to the sensation, as if you were a leaf blown along by the wind, until you realize that you are both the leaf and the wind. The world outside your skin is just as much you as the world inside: they move together inseparably, and at first you feel a little out of control because the world outside is so much vaster than the world inside. Yet you soon discover that you are able to go ahead with ordinary activities—to work and make decisions as ever, though somehow this is less of a drag. Your body is no longer a corpse which the ego has to animate and lug around. There is a feeling of the ground holding you up, and of hills lifting you when you climb them. Air breathes itself in and out of your lungs, and instead of looking and listening, light and sound come to you on their own. Eyes see and ears hear as wind blows and water flows. All space becomes your mind. Time carries you along like a river, but never flows out of the present: the more it goes, the more it stays, and you no longer have to fight or kill it.

Your Ego Is Not Your Friend

Suffice it to say that in order to achieve the state that Alan Watts achieved, you must overcome the efforts of your ego to keep it from happening. Therefore, it is imperative to realize, and to have in mind at all times, that your ego is a product of your mind. It developed as you grew from a child into an adult. and your beliefs and opinions were formed unconsciously, based on your environment and upbringing.

The upshot of knowing who you really are is this: If you don't like your personal reality, you have the power to do something about it, and in this regard, it's important to realize that beliefs—particularly the beliefs hidden in your ego, which tend to be unconscious—create your reality. This is so because belief—true, unadulterated belief—is powerful. The effectiveness of placebos, for example, has been demonstrated time and again in double blind, scientific tests. The placebo effect—the phenomenon of patients getting well or feeling better after taking dud pills—is seen throughout the field of medicine, and belief by a patient that he or she has taken real medicine is what causes it. One report says that after thousands of studies, hundreds of millions of prescriptions and tens of billions of dollars in sales, sugar pills are as effective at treating depression as antidepressants such as Prozac, Paxil and Zoloft. What's more, placebos cause profound changes in the same areas of the brain affected

by these medicines, according to this research. For anyone who may have been in doubt, this proves beyond a
doubt that thoughts and beliefs can and do produce
physical changes in our bodies.

In addition, the same research reports that placebos
often outperform the medicines they're up against. For
example, in a trial conducted in April 2002 comparing
the herbal remedy St. John's wort to Zoloft, St. John's
wort cured 24 percent of the depressed people who received it. Zoloft cured 25 percent, but the placebo cured
32 percent.

Taking what one believes to be real medicine sets up the
expectation of results, and what a person expects to happen
usually does happen. It has been confirmed, for example,
that in cultures where belief exists in voodoo or magic, people will actually die after being cursed by a shaman. Such a
curse has no power on an outsider who doesn't believe. The
expectation and belief causes the result.

Let me relate a real-life example of spontaneous healing that I believe came about because of her belief and
that of others. It involved a woman I'd known for quite
some time I will call Nancy, which is not her real name.

Nancy is a minister's wife. She's a devout Christian—
as firm a believer in her religion as a bushman who'd drop
dead from a witch doctor's curse is in his. Some years
ago, a lump more than half an inch in diameter was dis-

covered in one of her breasts. Her doctor scheduled a biopsy.

A prayer group gathered at her home the night before this procedure was to take place. Her friends prayed not that the lump would be benign, but rather, that it would disappear entirely.

Nancy is a member of a denomination that takes the Bible literally. In Matthew 18:19-20, Jesus is reported to have said, "Again, I tell you that if two of you on earth agree about anything you ask for, it will be done for you by my Father in heaven. For where two or three come together in my name, there am I with them."

As you can imagine, it was more than two or three. It was a living room full. The next morning, upon self-examination, the lump in her breast appeared to have vanished. But nonetheless Nancy kept her appointment at the hospital where her doctor conducted a thorough examination.

The lump indeed was gone. Not a trace could be found, and the bewildered doctor sent her home.

How could a solid lump of tissue disappear? It melted away due to the potent combination of belief and expectation. We indeed create our own reality.

Jesus also said, "Therefore I tell you, whatever you ask for in prayer, believe that you have received it, and it will be yours." (Mark 11:24) Notice the tense change in this verse. Jesus is saying to believe that you already have what you ask for and it will be given to you in the future. Jesus apparently knew that thoughts are things and that what we believe already exists in the nonphysical realm of spirit as a thought form. Thoughts are things, as we will see, that are ready to materialize on the physical plane.

How are beliefs able to do this? It has to do with the different levels of mind. You might call them lower and higher, or subjective and objective. What differentiates the higher from the lower is the recognition of self. Microbes, plants, worms, and fish possess the lower kind only. They are unaware of self. Even higher animals such as squirrels and other animals of the forest are likely totally unaware of self. This is indicated by the fact that an antelope, for example, does not seem to become angry with a lion when the lion kills and eats one of it's young. Once the lion is out of sight, the antelope simply resumes going about its business of grazing.

Perhaps some animals, dogs and other pets and perhaps dolphins, elephants and whales, have some level of self-awareness. I once had a dachshund that would let me know his displeasure by pooping on the rug when I left him alone for what he apparently considered too long a time. Certainly all humans, even small children, are self aware, and so it appears that the higher variety of self-aware thought is possessed in progressively larger amounts as if ascending a scale. At the present stage of evolution on Earth, humans possess the top level of consciousness, and in theory we each have access to all the levels, although in practice very few have trained themselves or been trained to access the higher levels. All of them are located outside the physical dimension, which is why I think of mind as being the universe's version of cloud computing.

Here are the levels of mind according to a professor at the College of Metaphysics in Windyville, Missouri:

1. An Individual's Conscious Mind
2. An Individual's Unconscious Mind, accumulated during the current life
3. An Individual's Subconscious Mind, accumulated during all his or her incarnations
4. The Collective Subconscious Mind of humankind, containing the archetypes and what is sometimes called the Akashic Records
5. The Subjective, Non-Dual Ground-of-Being Mind of the Source

Level Five, the ground of being subjective mind, is the organizing intelligence or mind present everywhere that, among other things, supports and controls the mechanics of life in every species and in every individual. It causes plants to grow toward the sun and to push roots into the soil. It causes hearts to beat and lungs to take in air. It controls all of the so-called involuntary functions of the body. And the fact is, it controls a lot more, including all physical and metaphysical laws.

Level Two, an individual's unconscious mind, contains the beliefs that have been established in this life, and like all levels after the conscious mind, the unconscious mind is subjective, meaning it cannot think outside of itself. This is why your beliefs create your reality. Your unconscious subjective mind determines your circumstances and your reality because it blends into and is part of the mind we all share and because of this influences events either favorably or unfavorably based on your beliefs. In addition, your beliefs —whether conscious or unconscious —influence the decisions and the choices you make.

Years ago, I read a series of lectures given in the early twentieth century in Scotland by a man named Thomas Troward (1847-1916) that made a lot of sense. He said the conscious mind has power over the unconscious subjective mind, and the subjective mind creates your reality. I discovered the truth of this firsthand in college when I learned to hypnotize others. I would put a willing classmate into a trance and tell him he was a chicken or a dog. Much to the amusement of my audience, he would then act accordingly.

Hypnotism works because the hypnotist bypasses his subject's conscious mind and speaks directly to the subject's subjective mind. Because of this, a subject's conscious, objective mind is unable to question or disregard the hypnotist's directive. Of course, once the subject emerges from the hypnotic trance, his or her objective mind will take over and will be able to nullify the hypnotist's directive. Nevertheless, while the subject remains in trance, the subject's subjective mind has no choice but to bring into reality what the hypnotist instructs it to do.

The Role of Feelings in Changing Beliefs

If you want to change a belief buried in your unconscious mind, it's important to realize that how you feel about the belief, or you sense of "knowing at a gut level" whether or not it is true, is as important as facts and logic are when it comes to convincing the subjective mind to discard it. So, if you have been brought up to believe and feel, for example, that you are a victim and will never amount to anything, or that people in your family are destined to be overweight, you actually will be a victim and never amount to anything, and you will also be overweight—until, that is, those beliefs change and your subjective mind is reprogrammed.

As stated, subjective mind cannot step outside of itself and take an objective look. As such, it is capable only of deductive reasoning, which is the kind that progresses from a cause (what is programmed into it) forward to its ultimate end. Having the mind of a deer, a rabbit or a squirrel, it does not stop to question or analyze. This is the same reasoning a criminal might use in committing a crime. He may walk into a room, see a man counting his money, and think: "I need money, so I will take his. Since the man is protecting the money, I will get rid of him. I'll shoot him. He'll drop to the floor. I will then take the money and run. I'll leave by the window." The subjective mind is non-dual. Right and wrong, good and bad, are never considered—only how to get to the end result.

On the other hand, the conscious mind, being objective and self-aware, can step outside. It can reason both deductively and inductively. To reason inductively is to move backward from result to cause. A police detective, for example, would arrive at the crime scene and begin reasoning backward in an attempt to tell how the crime was committed, and who might have done it.

So, if you have a victim mentality, your subjective mind will filter out all sorts of opportunities that come your way because it determines what you notice and are attracted to out of the literally millions of things you are exposed to each day, and it is determined to make your beliefs come true. Therefore, if your subjective is convinced you are a victim, and nothing you do can change that, it will dismiss out of hand all sorts of opportunities that might lead to a better life if you would only notice and take advantage of them.

It follows that if you want to change your life, you must change your beliefs, and this may not be easy. Repetition of the new belief you want to adopt will help, but it may not be enough by itself because, as mentioned above, you have to *feel* the belief you want to adopt is true.

Toward Higher States of Consciousness

Many people nowadays believe that have evolved on Earth because of an innate desire to become perfectly balanced in terms of love and wisdom until we finally merge back into the source at the end of our journey in the physical plane of existence. This means that to move ahead we must find the "distortions" within us—what I think of has harmful beliefs or negative thought structures. Some would say they are "shadows." Whatever term you prefer, you must purge them from your unconscious mind in order to advance.

Since the unconscious, subjective mind does not know what are good beliefs and what are harmful beliefs, and you may not even be aware they are buried there, the first step is to identify the beliefs you need change. Beliefs are points of view you have about yourself and the world, and points of view can be changed. They are ideas we think about often, and once we have thought frequently enough about one, the unconscious mind assumes it must be important to your survival as the person you think yourself and wish to be. So, the mind condenses it into a belief, once an idea has become a belief, it exists outside of your conscious awareness.

Let's say because of how you were raised you have a belief you are unworthy of love. If someone asked you if you think you are unworthy of love, you likely would say "No, of course not." You might even be insulted the individual had the nerve to ask you that.

Unfortunately, you cannot simply look into your conscious mind to discover your beliefs. The way to go about it is to observe your own behavior and life situation because how you behave and where you are today are the results of your unconscious beliefs and opinions. As already discussed, in addition to influencing how you react and your choices, your unconscious mind blends with the universal subconscious mind, causing it to bring into your life what you unconsciously believe about yourself and the world outside you.

Let me interject here that a frequent misconception is that your thoughts create your reality. This is not so because thoughts occur only in the conscious mind, and the unconscious mind is the instrument that creates your reality. Only when a thought is given enough significance to create a belief does the thought gain power. Therefore, a thought you consider to be insignificant or untrue is powerless. The truth is, you give all thoughts their meaning, or the lack thereof. As has been said, a subjective mind does not judge whether something is right or wrong, good or evil. You arrived on this planet a blank canvas on which you can paint the life you want, so why not take advantage of that and do so?

As has been said several times, it is important to understand that your ego is not you and does not have your best interests at heart—that the ego's primary goal is its own survival. It is fully aware of your personal reality at all times and constantly judges whether your reality is in sync with your unconscious beliefs. It draws upon them to create thoughts and impulses that attract your attention, thereby seeking to reinforce and uphold those beliefs. Therefore, unless you discover what your unconscious beliefs are, the ego will do its best to be in charge, directing your life.

You can determine the beliefs that govern your life by noticing what repeatedly happens in your life. For example, are you a man or a woman who always seems to attract a member of the opposite sex who ends up abusing you? If so, you must have an unconscious belief you are unworthy and deserve abuse. What are other things that consistently go wrong? A buried belief is the reason.

Something else you can do is take note of anything that sets you off or "triggers" you because that is a result of one of your core beliefs. Here's a simple example: suppose your father was critical of you and frequently criticized you with the result as you grew up you felt that nothing you did was ever good enough. Now let's say in your current job you have a male boss older than you, and any time he gives you criticism—even though it may be constructive and meant to be helpful—you immediately feel a sense of fear and anxiety. You have just experienced a trigger because you were conditioned by your father from childhood to believe you are incapable of success. The same process is playing out in obvious and subtle ways in every aspect of your life.

If you want to improve your life, you need to notice any time you react to a situation. Then think about your reaction and drill down until you identify the core negative belief that brought it about. Once you have put your finger on it, you can begin the process of changing it to what you want to believe about yourself. Once the belief has been expunged, or replaced with one that's positive, your life will change for the better. Some say the energy centers of our body develop blockages based upon such thoughts and perceptions. If one is lacking in some form of self-love or wisdom, that distorted energy will manifest in one or more of the energy centers. In order to unblock these energy centers to become balanced, we must heal the distortions within us.

I'm aware of two ways this can be done, which are known as the "feminine" approach and the "masculine" approach, the feminine being the positive and masculine the negative polarity—the positive being love and the negative being wisdom. This does not mean that love is good and wisdom is evil. Both polarities are equally valid expressions of the Source, and that achieving the balance between love and wisdom within the self is crucial for spiritual evolution. Both approaches, while different, can be effective. The feminine heals through transmutation, and the masculine heals through recognition. Which one you favor is up to you.

The Feminine Approach to Balancing

The feminine approach to balancing is accomplished through feeling thereby purging the unwanted thought form through love. It involves going to the root of the negative emotion, feeling it completely, and allowing it to express itself. Rather than meeting it with resistance, the key is to meet it with love and acceptance, and then, although it may be painful, review and to an extent relive the experience that caused it.

The feminine approach is much more painful and intense than the masculine approach, but it is also much more powerful and immediate. A single healing session has the potential to purify an old wound that has been festering for decades. It takes time, practice, and courage to develop the skill of locating repressed emotions and the memories that are their causes so that you can face them and feel them, but the power of the distorted beliefs and points of view they brought about will be healed quickly as a result.

The Masculine Approach to Balancing

Attaining wisdom and understanding is the masculine approach to healing. Illusions are banished by perceiving that that is simply what they are—illusions. Unlike the feminine, which seeks to heal repressed energy by feeling it out of existence, the masculine approach seeks to defuse and disperse the energy out of existence. It is not as immediate and powerful as the feminine approach, but it is less painful and intense. It is the one I personally prefer, and the one that works best for me. What is required is a permanent shift in perception or attitude, which over time will drain a debilitating thought form of its power.

Here is how I suggest you begin. Step back and observe your own behavior. Once you become aware what is going on, and you are conscious of how you are reacting in different situations, you will have the power to change. To employ this technique successfully, my advice is to put all your energy and passion into your desire for change. You must want the new belief to be true with your entire being. A fraction of a second exists between the moment something happens and your "triggered" reaction. Stop yourself at that instant. Then change what you think and how you react to conform to your new beliefs—even if at first this feels a little awkward. This will take energy away from the old beliefs and direct it onto the new ones, and it won't be long before you start to feel the truth of your new beliefs and comfortable being your new self.

This is how you change beliefs using the masculine approach: you see them for what they are and *want* them to be. Changing beliefs using this technique takes effort because the unconscious mind needs to see proof it is safe to let go of something that has always been seen as beneficial to your survival. You can do this by constantly and consistently presenting your unconscious mind with the truth and by putting some a effort behind it. By acting as though your new way of reacting is who you are now, your unconscious mind will get the message, and eventually, the desired action will feel natural and come naturally.

Let's get into this a little more deeply. Some people think that they cannot change their beliefs because they cannot get themselves to feel the desired belief is actually true. Even though intellectually and logically they may see an old belief as false, it still feels true, and so they think they are powerless to change. This is because beliefs lead to points of view that you possess, and it takes conscious effort to change a point of view. This is likely why you cannot *feel* what you would like to feel about the new belief you want to adopt. The old points of view are still there at the unconscious level, hanging on, preventing you from releasing the old belief. To let one go, you must admit to yourself, and perhaps even to others you have had conversations with about whatever it is, that you have been wrong—you have been guilty of an error in judgment. In other words, you must identify the points of view and opinions a belief has created and adopt new points of view based on the new belief—thereby replacing the old.

I am living proof that if you really want to, you can change your points of view, and therefore your beliefs through wisdom, knowledge, and your fervent desire to live the truth. As a young man, I was an agnostic bordering on atheist and a confirmed Scientific Materialist. I thought that when you died, that was it, so why not seek pleasure as your number one goal in life. Not so now by any stretch of the imagination. Have you ever seen *Animal House?* That was me back then.

If this approach doesn't work for you at first, it is because some part of you still isn't completely sure that balance and total empowerment is really what you want. In order to change, you must lose all interest in your old beliefs and the old stories that created them. Think of yourself as the captain of a ship, and you have turned the ship in the direction you want to go. The wake is your past—it's behind you. Leave it there—forget about it. It's gone and will dissipate and dissolve into the ocean.

The truth is, you will never break free from the grip of a belief and point of view until you let it go and forgive whoever and whatever brought it about. That's right—

forgive and forget. Uncle Charlie molested you and that's why you fear sex and despise men? Uncle Charlie was a pathetic, dirty old man nobody loved, who is worthy of pity. Forgive him.

Mom said you were lazy and good for nothing? She was angry, she was wrong, and her father was a loser who drank himself into the grave. Forgive her. Holding on to bitterness isn't hurting Uncle Charlie or your mom. It is only holding you back and making you miserable. As soon as you desire empowerment more than anything— even more than revenge or sympathy—as soon as you truly make the effort, cultivate the desire, and jettison old baggage, you will be on your way. Then nothing can stop you. Claim it whether or not it feels true right now. Claim it because you want it to be your reality, and you want to be totally free.

When something in you starts to resonate with your new belief, you will have begun the process of implanting it in your unconscious mind. You must send a feeling-based message to that part of you. By claiming it because you truly want it, that is what you will be able to do.

Chapter Five: Clean Out Your Attic

Give this some thought. You have been around and evolving for a long time—all the way back to the first single-cell creatures that formed in the primordial sea. You are the product of evolution that took place over a mind-boggling 3.77 billion years. Other hominoids that evolved along the way branched off onto dead-end paths or developed into chimpanzees or gorillas and such. Some, like the Neanderthals, made it pretty far along the evolutionary path, but eventually could no longer hack it and became extinct when your ancestors, Homo sapiens, came along and took over their territory. But you kept going and going like the Energizer Bunny. You continued evolving until you eventually arrived at the very top of the food chain. There can be no doubt about it at all. You are a member of a very exclusive club—one among the most gifted and highly intelligent creatures that has ever lived. You are the pinnacle of life on earth, and you are now ready to make the leap into the kingdom of heaven.

You can do this because your mind is fantastic. Scientists say we humans typically use only a portion of its capacity. It is the most important tool you have—an amazing tool that is very much like a garden. You can cultivate it, pull out the weeds, water it, plant the right seeds, and allow them to grow. Or you can neglect it, and let it run wild. Either way, cultivated or not, it must and it will bring forth whatever is allowed to grow in it. There may be ideas and beliefs that were planted in your mind while you were going up that are doing you no good. They are weeds that need to be pulled out and cast aside. Plant good, helpful ideas and beliefs in their place. If good seeds are not planted in your mind and allowed to flourish, destructive weeds will take over and will continue producing more of their kind. Just as a gardener or farmer cultivates his plot of land, weeding it, and growing flowers and the fruits and the vegetables he or she wants on the dinner table, so may a person tend his or her mind, weeding out the useless and destructive thoughts and cultivating only those that have the promise of bearing delicious fruit. If such cultivation does not take place, if discipline is not exercised, the result will not be good because what's in your mind will eventually be what's in your world.

As was discussed, over time the outer conditions of a person's life always come to be in tune with his or her inner state. By the process of planting and cultivating positive, constructive thoughts, you will sooner or later discover that you are the master gardener of your mind— the director of your life; captain of your ship of fate. You will also come to understand that your thoughts and beliefs shape your character, which also creates your circumstances. Ultimately, what's in your mind is your destiny. James Allen [1864-1912] wrote a book about this that I highly recommend called, *As a Man Thinketh*.

Something else to know along this line is that the unconscious mind does not understand, or perhaps simply doesn't hear the words "no" and "not." Suppose, for example, you're a tennis player. You're in a big match, it's close, and you are now in a tiebreaker. The next point could decide the match. The point has come for your opponent to serve. As you pass by him at the net changing ends, you say, "You're playing great today, Henry. Don't blow it. This is a big point coming up. Whatever you do, do not double fault."

You've started Henry worrying, and on top of that, his unconscious mind doesn't hear or understand the word "not." All it hears is "do double fault," and it takes that as a directive. Try as he might to do otherwise, Henry will double fault.

Actually, I advise you not to play such a dirty trick. As the saying goes, "What goes around, comes around," and you don't want that sort of negative behavior coming back at you. More will be written about this. The important thing to remember is that self-talk and coaching should always be framed in a positive way.

Think Positive and Rid Yourself of Fear

Let's dig into the issue of fear because fears are beliefs and feelings, and beliefs and feelings are what create your reality. To learn what you fear, as suggested in the previous chapter, tune into your moment-to-moment stream of consciousness and observe what makes you worried, anxious, resentful, uptight, afraid, angry, and so on. Step outside yourself and identify unsettled emotions, tugs and urges that have become part of your programming. Slow down and consider what triggered a negative emotion. Did your temper flare? Why? Why was it so important for things to go a certain way? If you trace what you felt back to its cause, you might come to a particular variety of fear, and it's been said that only two fears are instinctive: the fear of falling and loud noises. Other fears were acquired, and whatever was acquired can be disposed of.

According to some experts, the fears that hold people back can be grouped under one of six headings:

1. the fear of poverty (or failure),
2. the fear of criticism,
3. the fear of ill health,
4. the fear of the loss of love,
5. the fear of old age,
6. and the fear of death.

I've listed the fear of poverty (failure) first because in many ways it can be the most debilitating. It is self-fulfilling in that traits develop that bring it about. For example, are you a procrastinator? An underlying fear of failure is probably the root cause and can be counted upon to produce the result you fear.

Are you overly cautious? Do you see the negative side of every circumstance or stall for the "right time" before taking action? Do you worry (that things will not work out), have doubts (generally expressed by excuses or apologies about why one probably won't be able to perform), suffer from indecision (which leads to someone else, or circumstances, making the decision for you)?

Are you indifferent? This generally manifests as laziness or a lack of initiative, enthusiasm or self-control.

Step back and listen for internal voices that say "can't" or "don't" or "won't" or "too risky" or "why bother?"

How do you get rid of them? Shoo them away.

Whether you are the president of a company, or a bum on Skid Row, the only thing over which you have absolute control is your thoughts.

You may say, I can't control what thoughts pop into my head. True. You may not control what thoughts arise, but you can decide whether to discard one or to keep it. You can decide that it is counterproductive and throw it away, or you can turn it over and over in your mind, nurture it and let it grow. Whatever thoughts you keep will expand and eventually manifest themselves.

Beginning now, each time you catch yourself with a negative thought, a thought that says "you can't," "it's not possible," "maybe someone else but not me," get rid of it. Shoo it away.

But you say, "I'm poor, I'm not a good student, I'm not a good salesperson, I'm in the lower third of productivity."

That's your ego talking. You are what you are because of your unconscious beliefs. You want the best for yourself, but your unconscious ego-mind is holding you back because of the way it was programmed.

If what I've been writing about on this page is a serious problem for you, follow the advice given in the last chapter, and go out and buy some self-help tapes that will plant positive thoughts in your mind in place of the negative ones. Play them to and from work and before you go to sleep at night. Use self-hypnosis tapes. Play them over and over for at least a month. Get all that junk out of your head, and replace it with thoughts that are positive.

What about the other fears? They're to be discarded in the same manner. If you suffer from fear of criticism, for example, it probably came about as a result of a parent or sibling who constantly tore you down to build himself up. You'll know this is a problem if you are overly worried about what others might think, if you lack poise, are self-consciousness or extravagant. (Why extravagant? Because of the voice which says you need to keep up with the Joneses.) You must rid yourself of inner voices that tell you to think even twice about what others will say. Take advantage of that fraction of a second between stimulus and response, stop to remember this, and let it sink in. That's how you can change2 how to eliminate destructive fears and beliefs.

Let's think for a minute about the fear of criticism. There have been places and times in history when what others thought was worth worrying about. My great, great, great, great, great, great, great grandmother, Suzanna Martin, for example, was accused of being a witch, falsely convicted, and hanged in Salem, Massachusetts, in 1692. She was an old lady. Probably, she looked like a witch. But her downfall was the stir she caused after her husband died. She was able to run the farm successfully without a man around. Think of the talk. Such a thing wasn't possible, or so they believed, without the use of witchcraft.

The opinions of Suzanna Martin's neighbors mattered a great deal. They led to an unpleasant and untimely death.

What about today?

In Iran, China, Russia, or North Korea one might have to watch out what neighbors think or what the "virtue police" hear, but this simply is no longer a valid concern in developed, democratic countries. What others think or don't think of you or anyone else is their problem. Yet worrying about what they think can cause a great deal of misery, create karma that will have to be worked out, and cause interference between your conscious and your subconscious minds that blocks the channel of communication.

What about the fear of ill health?

To rid yourself of this, it should be enough to know that what you worry and think about is eventually what happens. Ever noticed that it's the people who talk about illness, worry about illness, are preoccupied with this or that possible illness, think they feel a pain here or there or were exposed to some germ, who are precisely the people who stay sick most of the time? The power of suggestion is at work.

How about the fear of the loss of love? This one manifests in the form of jealousy and is self-fulfilling like the others. The person you try so hard to hang onto feels smothered, with the result that you end up pushing that person away. Try being yourself. Give them love, but give them room. It they leave you, they would have done so anyway. You can now move on to a truly meaningful relationship.

Next is the fear of old age. This is closely connected to the fear of ill health and the fear of poverty because these are the conditions a person really is concerned about deep down. The power of suggestion is hard at work here, too. If you think you're too old to do this or that, you will indeed be too old.

Consider this. My children are the same flesh and blood as my wife and me. I saw them being born, still connected by umbilical cords. I clipped the cord of one of them myself. My wife was thirty-six at the time our youngest was born. I was fifty-four. Yet the cells in my body, and in my wife's body, and in my son's body all were the most recent in an unbroken chain of cell division that goes back to the first life on earth. All the cells—my wife's, my son's, and mine—are at the end of a chain that is precisely the same age: billions of years. As the physic Edgar Cayce often said in trance, "Spirit is the life, mind is the builder. The physical is the result." Those telomeres get shorter and shorter because you think that you should look and feel older as the years go by. The physical body is the overcoat of the mental body. It gets old and decrepit because a person expects it to. It gets older and decrepit because a person stops learning, growing, and playing a role in the evolution of humankind.

When you've learned all you can from this life, the time will come for you to check out. And check out is what you will. No one says you have to be old.

Now we've come to that final bugaboo, the fear of death. As you now have seen, there's nothing to fear except having been fearful in this life. Consider the millions who have had near-death experiences and are no longer afraid to die. They're convinced they'll be greeted by their guides as well as by loved ones who have gone before. They look forward to being bathed once again in the all-encompassing light, which many have described as total, unconditional love. Most do not expect to experience pain. It has been reported by many that the spirit exits the body the instant it looks as though death is inevitable.

Only a handful who have had hellish experiences worry about what they may encounter in the nonphysical world. What they need to realize is that each of us creates his own reality. We experience what we expect to experience, what we think we deserve. In the physical world, this takes time. In the nonphysical world of spirit, which is the medium of the mind, we instantly create our reality, just as we do in dreams. If we expect Hell, the Hell we believe we deserve is the Hell we will get. If we expect Heaven, our vision of Heaven is what we will have.

Anyone who has ever had a lucid dream will understand what I mean. Such a dream is one in which a person realizes he's dreaming. I've had many and I look forward to them because it's more fun than Disney World. As soon as you're aware you're dreaming, you can begin to compose the dream, determine the players, the surroundings, the action. Want to fly over the Grand Canyon? All you have to do is "think" this. Fly over is what you will do, no airplane required. Like anything it takes practice, but I've gotten so I can swoop and turn and loop the loop.

Want to attend a cocktail party populated by Hollywood stars? You'll be there with Robert DeNiro or Julia Roberts. These characters will, of course, be your own thought projections.

You are a dreamer in the Creator's big dream of life, and you can make your waking dream lucid as well. Until now, you may have thought you were at the mercy of conditions outside yourself, that you've either been lucky or unlucky, that chance has brought you where you are. This isn't true. You've brought yourself to this spot, either consciously or unconsciously. If this is not where you want to be, you've arrived because your unconscious mind has been programmed incorrectly, and you are totally out of touch with your Higher Self. Perhaps you hear snippets from it every now and then but ignore what it's trying to say because of other voices which beat it back with "can't," "don't," "shouldn't," "too risky." These are the words of your ego. Your Higher Self wants you to evolve and to enter the kingdom.

Until you started reading this book, you may have thought you were at the mercy of conditions outside yourself, that you have either been lucky or unlucky, and that chance has brought you where you are today. No so. You brought yourself to this place, and you did so unconsciously. If this is not where you want to be, you will have to change your programming, which likely took place when you were a child.

The Power of Positive Thinking

I suspect you have heard about "The Power of Positive Thinking," that positive thoughts are much more likely to produce good results than negative thoughts. I'm reminded of *The Little Engine That Could,* an American fairy tale published numerous times in illustrated children's books and movies since its original debut in 1930. The Little Engine was a railroad locomotive that was tasked with pulling a long, heavy train—one that seemed much too large for it—up and over a mountain. But even so, the Little Engine was determined and kept telling itself over and over, "I think I can, I think I can." It was a struggle, but the Little Engine persevered and finally succeeded.

It's a good story and a valuable one to teach young children the benefits of optimism and hard work. The problem is, many of us today were not taught that lesson as children, and in fact, feelings of frustration, discontent, and dissatisfaction were ways of solving problems that many of us "learned" as infants. For example, if a baby is hungry, he or she expresses discontent by crying. Lo and behold, a warm and tender hand appears magically out of nowhere and brings a bottle of milk. Later on, if the baby is uncomfortable, again, he or she will again express dissatisfaction, and the same warm, comforting hands magically appear and solve the problem. That's fine for babies but, unfortunately, many children continue to get their way and have their problems solved by indulgent parents merely by continuing to express their feelings of frustration when things don't work the way they want. All they have to do is feel frustrated and dissatisfied, express their dissatisfaction, and the problem will be solved. Sometimes what have become known as "helicopter parents" continue to cater to their children in this way all the way through high school, college, and beyond.

This way of life "works" for infants, and for some children. But it does not work in adult life when a person is out in the world on his or her own. Yet many continue to expect, perhaps unconsciously, that it will work. They seem to think that by feeling discontented and expressing their grievances—if only they feel put upon enough—life, or someone will take pity on them, rush to their aid and solve their problems. Let me assure you that 99.99 percent of the time that is not going to happen. It is my advice that you take responsibility for every aspect of your life.

Imagine, for example, you land an entry-level job as a management trainee in a big corporation. With you in training are several other bright young men and women fresh out of business school. Imagine the way things work in this company is often not to your liking. Management trainees, for example, are relegated to cubicles with five-foot-high walls affording little or no privacy, while the senior staff all have corner offices with large windows and spectacular views of the East River. You spend a good deal of time grumbling to yourself and to others about this injustice, subconsciously believing that will get you out of that cubicle and into a corner suite. Your fellow trainees, on the other hand, spend their time making positive suggestions and anticipating and providing for the needs of customers as well as for fellow workers higher up on the corporate ladder. Whom do you suppose is most likely to be first to break out of his or her cubicle? The one who constantly complained? Or the one that consistently delivered the goods?

Don't you feel a twinge inside that intuitively "knows" the positive attitude, the attitude of service to others, will inevitably win the day? That "twinge" is a message from something inside you that knows the correct answer called "intuition."

If you have been ignoring that feeling when it comes, now is the time for you to begin recognizing such messages. They have a light and airy feeling to them, even though they may seem to run counter to egocentric notions, such as, "The first order of business is to look out for number one." That egocentric notion may work in the short term, but in the long term, it is bad advice. The fact is that it's always best to under-promise and over-deliver to customers and bosses—as well as to anyone else for that matter. By over-delivering, your reputation grows as you create positive vibes and positive opinions of you by those with whom you come in contact. A reputation that you are someone who can be counted on can only lead to good outcomes and opportunities for you in the long run.

Let's consider for a moment why some people may spend their valuable time on earth grumbling and complaining away opportunities to get ahead. It's often because they have felt frustrated and defeated for so long—ever since they were babies in a crib and while growing up with indulgent parents—that those feelings have become ingrained. Their minds are in a kind of holding pattern, and it's never occurred to them to step outside of themselves in order to get in touch with intuition that would tell them, if they would only listen, that grumbling and complaining are counterproductive and accomplish nothing. Until they wise up, they will continue—to their own detriment—to radiate those feelings, and as sure as night follows day, their discontent will lead to failure.

No matter what your mindset, if you want to change it, it's important to realize, as has been discussed, that beliefs and feelings are intertwined. It might be said that feelings are the soil in which thoughts and ideas grow. If you are habitually grumpy and in a bad mood, you need to lighten up and begin seeing the glass half full. Moreover, when you begin working toward a goal, try thinking how you will feel when you reach it—and then actually make yourself feel that way. I'm serious. Conjure up the feeling of "Success!" The thrill of accomplishment will communicate the belief to your unconscious mind that it's inevitable you are going to achieve what you have set out to accomplish. The feeling creates the belief, and the belief creates the feeling. A mental model of success will be etched into your unconscious, and that the desired outcome will surely come about.

Let's say you are pursuing a challenge and fervently want to accomplish it. Assuming you have the education, the knowledge, and the qualifications required to reach your goal, and assuming you feel strongly about it at an emotional level, you will almost certainly realize success. This will happen because your unconscious mind, which is connected to the universal subconscious and all other minds, will go to work and act like a magnet, drawing to you what you need. The greater your desire, the more powerfully your unconscious will mind strive to produce results.

In summary, belief and emotions are the keys. It's important to feel the joy of having accomplished what you set out to accomplish before it actually happens. This will convince your subconscious the goal has already been reached and the universal mind will cause it to be reached as a result.

Be Likable and Appealing to Others

Leaving psychological suffering behind will require effort and work, and that means things will be easier if you have help along the way. The most likely source of that help will be friends, partners, and mentors. Obviously, the going will be easier and you will attract more help if people like you and want to work with you. Therefore, it should go without saying that it's important to be someone others want to be around—someone people would like and want as a friend. That means you need to be someone who "talks the talk," and "walks the walk." Perhaps you know a person who does the opposite. If not, you are likely to come across someone like that in your business dealings, so be prepared and never, ever be one of them. In public such people talk openly—some even brag and boast—about the importance of having integrity and doing the right thing. But in private it's a different story. They bad mouth people and do things that aren't consistent with the honest-John public persona they hope to project. People quickly see through these phonies. As the old saying goes, "Say what you mean, and mean what you say," and people will respect you for doing so.

Obey the Laws of Physics

If you are a Materialist and think matter is all that exists—that there is no God and nothing spiritual—you might come to the conclusion you can do whatever you want, harm whomever you want, and never have to suffer any consequences. But that is not how things work. I agree it may work for a while, but eventually you'll get back what you gave out because, just has there are laws of physics, there are laws of metaphysics. People must obey them, and nations must as well. Take Japan and Germany in World War II as examples. Both countries had astonishing victories in the beginning. Each country benefited from a national zeitgeist that they were invincible. But they ended up being crushed because the atrocities they committed came back upon them with a vengeance.

More will be written about this in the final chapter of this book.

Always Keep Your Life in Balance

You are probably familiar with the ancient Chinese symbol composed of a white "yin" interlocking a black "yang" that represents dual nature of things. It symbolizes that we live in a world that is composed of opposites: Up, down, black, white, good and evil. Without the tension opposites create, nothing would or could exist—everything would fall apart. Follow the advice of this book and enter the kingdom but do not allow complacency to set in. Always seek new challenges, realizing that without one, self-destruction may result. Always strive to continue growing.

It can also be comforting to know there can be no growth without at least some discontent. Deep within, you know what is best for you. There is an urge built into you that pushes you to strive for growth, and for most of us, growth will not continue without some agitation and discontent. So study your dissatisfactions. They will tell you what you are about to leave behind and possibly point you in a new direction. Be willing to be uncomfortable. It is the way to grow.

As you contemplate your future course, it is also important to realize you can only attract that which you feel worthy of. Self-esteem is critical to success. That's why I urge you to get rid of the psychological baggage. The truth is you are not what you have, and you are not what you do. Beneath your fear and negative programming, you are perfect—an enlightened soul, fully self-actualized and a living example of unconditional love. The more you can let go of fears, the higher your self esteem will be, and the more options you will have and more risks you can take. The more you like yourself, the more others will like you, and the more worthy you will feel.

You can have anything you want if you can give up

the belief that you cannot have it—assuming what you want does not conflict with someone else's belief. If, for example, you desire a fulfilling, one-to-one relationship, but demand it to be with a particular person, you are not operating in harmony with the universe.

Another example is in the area of accomplishment. You must get the education necessary to create what you want. "Where your attention goes, your energy flows." You attract what you believe you are and that which you concentrate upon. If you are negative, you draw in and experience negativity. If you are loving, you draw in and experience love. You can attract to you only those qualities you possess. So, if you want peace and harmony in your life, you must become peaceful and harmonious.

Something else to understand is that a stronger emotion will always dominate a weaker one. Every idea can be the beginning of a manifestation—although unless you nurture it, think about and develop feelings about the idea, it will not become expressed in reality. It does not matter which idea you consciously favor, even know to be desirable, a stronger emotion will nullify a weaker one, and the strongest emotion will begin to permeate all aspects of your activities. For example, if you are emotionally focused on the sexual desirability of a particular person, you may begin to create circumstances that will increase the likelihood of an eventual sexual union.

It is also important to realize that new information you accept into your mind will destroy previous information of a similar nature. Once a pathway of information has been created in you, a new viewpoint will develop and prevail unless new information comes in to replace and destroy it. Let's say, for example, you fall off and get hurt while horseback riding. That may be the end of your experience with horses because you will have just been programmed negatively about horseback riding. This is why instructors always urge new riders to climb back aboard immediately. You need new, positive input to erase the trauma of the fall.

The mind is engaged in an endless state of growth and reorganization. As a result, it is possible to reprogram yourself. You can do this by using the feminine or masculine techniques described previously, or some combination of the two, and by reinforcing new beliefs and points of view by repeatedly listening to success-meditation recordings and using with visualization techniques. If you feel anxiety in crowds, imagine yourself relaxed in a crowd of people. When you fear doing something, and yet have the courage to do it anyway, you will soon do a mental flip-flop and may even become addicted to doing it.

Here is a case in point. Suppose you fear skydiving, or skiing fast almost straight down a steep mountain. If you force yourself to do so anyway, the experience will release endorphins, which are produced by the central nervous system and the pituitary gland and can produce a feeling of euphoria very similar to that produced by opiates. The result can be that you become somewhat addicted to skiing fast down mountains and skydiving.

You have within you everything required to make your earthly incarnation a paradise if you choose to accept that which is your divine birthright. We live in a universe of abundance, although the majority of humans populating our planet appear to view it as a universe of scarcity.

Heed the call. Take the leap. But do not go off half-cocked. Plan it out. Take a full day. Take more than one. Take as many as necessary to develop your plan.

Parting Thoughts

At the risk of repeating some thoughts and advice I've given in other books, let me leave you with this:

- Take five minutes, twice a day to affirm your goals, dreams, and desires. Most of us do not achieve our goals, not because we are too lazy or untalented, but because we forget about them and focus our efforts elsewhere.
- Spend some time in nature. Even if it is for just ten minutes a day, take the time to go for a short walk or sit in a place surrounded by nature. Release the stress of the day by com-

muning with God's creation, and you will soon feel recharged.

- Exercise. Your body is your temple. Take care of your temple every day. If your body is not in top form, neither will you be. Exercising, eating healthy and taking care of your amazing vehicle in this reality is a requirement for you to be able to produce at the highest levels.

- Meditate. The biggest improvements in our lives come from within. An effective way to release the limiting beliefs and destructive thoughts that may plague you is to meditate for thirty minutes a day. Among the many benefits, meditation teaches you focus, and opening to the transcendental part of yourself is strongly affected by your ability to focus. Regular practice of meditation has been scientifically proven to change your brain chemistry, lower blood pressure, make you sleep better, feel less stressed.

- Smile a lot. A smile can change the world. Not only for you but also for the people you interact with. Practice a genuine smile and give joy to the world. Impact the world today by smiling at everyone around you.

Find more ways to have fun. Life does not have to be a strict, gloomy experience we struggle through. Instead it can be full of amazing twists and turns. Think of it as an adventure because, as you now know, that's what it is. Approach it as such.

- Make sure you laugh out loud at least once a day. Do something stupid, childish, and completely weird. Be yourself, have fun, laugh at your own jokes.

- Remember that everything begins as a thought or idea. Ideas and experiences create beliefs that in turn, create your reality. If you are unhappy with your current reality, you must change your beliefs and your behavior. Beliefs should be changed when you recognize which ones are not working for you. Change that belief, and your life will change.

- Take a full day each month to ask and answer questions and develop goals and plans that will lead to a better and more blissful life.

- And finally, seek harmony in all that you do.

# # #

Appendix: Chapter One from the book, "Life After Death, Powerful Evidence You will Never Die"

Consciousness and the Brain

What if you knew you would exist forever? Would it change your outlook? Would you do anything differently? Would it make you happy? Sad? Uneasy? Would it cause you to want to work on yourself to become someone you would not mind being with forever? I think it's true—you are eternal. Why do I think so? In my book, *Amazing Truth*, I present the findings of quantum physics experiments that any fair-minded person would have to agree is powerful evidence consciousness is the ground of being of reality. Your consciousness, and that ground-of-being consciousness, are one and the same. All of us, and everything else, arose from it. Our consciousness does not result from electrons jumping across synapses in the brain. The brain is merely a receiver that connects our consciousness to our bodies. As a result, we are eternal.

To begin to convince you of this, I'm going to start by summarizing a lecture recorded on video given in India in 2011 by Bruce Greyson, M.D., The Chester Carlson Professor of Psychiatry and Director of the Division of Perceptual Studies at the University of Virginia. His job is to study consciousness and what he has to say about it may open your eyes to the truth.

The bottom line takeaway of Dr. Greyson's lecture is that brains do not actually create consciousness, despite what many scientists still think. He does say, however, that this mistaken belief is understandable since evidence does exist that the brain produces consciousness. Consider what happens when a person drinks too much or gets knocked on the head. Also, it's possible to measure electrical activity in the brain during certain kinds of mental tasks and to identify correlations between different areas of the brain and the different activities. We can stimulate different parts of the brain and record what experiences result, and we can remove parts of the brain and observe the results on behavior. This suggests that the brain is involved with thinking, perception, and memory, but according to Dr. Greyson, it does not necessarily suggest the brain causes those thoughts, perceptions, and memories. What the measurements actually show are correlations, rather than causation. The truth is that thoughts, perceptions, and memories, actually occur somewhere else and then are received and processed by the brain in a way similar to how a television, cell phone, or radio receiver works.

Western science, Dr. Greyson pointed out, is largely reductionist. It breaks everything down to its component parts, which are much easier to study than the whole, but the component parts do not always act like the whole. The brain is composed of millions of nerve cells or neurons, but a single neuron cannot formulate a thought, cannot feel angry or cold. It appears that brains can think and feel, but brain cells cannot. No one knows how many neurons are needed in order for them to collectively formulate a thought, nor do we know how a collection of neurons can think when a single neuron cannot.

Scientists get around this by saying consciousness is an emergent property of brains, a property that emerges when a large enough mass of brain cells gets together. According to Dr. Greyson, however, saying something is an emergent property is a way of saying it is a mystery that cannot be explained. It is a fact that there is no known mechanism in the brain or anywhere else that can produce non-physical things like thoughts, memories, or perceptions. The materialistic understanding of the world fails to deal with how electrical impulses, or a chemical trigger in the brain, can produce a thought or a feeling, or for that matter, anything the mind does. Despite this, according to Dr. Greyson, most scientists continue to maintain what he labeled, "The nineteenth century, materialist view that the brain in some miraculous way we do not understand produces consciousness."

These scientists, he said, "Discount or ignore that consciousness in extreme circumstances can function very well without a brain."

Dr. Greyson noted that the idea the mind and the brain are separate is what most people believed until a couple of hundred years ago, but in the nineteenth century western world, beginning with the Darwinians, science began exploring the idea that the physical brain might be the source of thoughts and consciousness. Ironically, as one group of scientists attempted to explain consciousness in terms of Newtonian physics, scientists in a different discipline, physics, were forced to move away from Newtonian physics and develop quantum mechanics in order to explain phenomena in which consciousness—what a researcher knows or doesn't know—completely changes the results of certain experiments. It is as though the right hand did not know what the left hand was up to. Incredibly, this remains how things are today.

Dr. Greyson listed a number of examples in his lecture of evidence researchers with the Division of Perceptual Studies—established in 1967 at the University of Virginia—have collected that demonstrate that consciousness can exist without a brain being involved. It is a testament to the stubbornness of materialist scientists that even though Dr. Greyson and his colleagues have

been collecting this data for fifty years, and many papers and books have been written and published revealing a great deal of it, most western scientists are unaware of this evidence. As a result, you will soon have a leg up on many western scientists.

The evidence falls into four categories:

1. Recovery of lost consciousness in the moments or days prior to death among people who have been unconscious for prolonged periods of time.
2. Complex consciousness ability in some people who have minimal brain tissue.
3. Complex consciousness in near-death experiences when the brain is not functioning or is functioning at a greatly diminished level.
4. Memories, particularly among young children, accurately recalling details of a past life.

Deathbed recovery of lost consciousness

The unexpected return of mental clarity shortly before death by patients suffering from neurological or psychiatric disorders has been reported in western medical literature for more than 250 years. There are published cases in the medical literature of patients suffering from brain abscesses, tumors, strokes, meningitis, Alzheimer's disease, schizophrenia, and mood disorders, all of whom long before had lost the ability to think or communicate. In many of these cases, evidence from brain scans or autopsies showed their brains had deteriorated to an irreversible degree, and yet in all of them, mental clarity returned in the last minutes, hours, and sometimes days before the patients' deaths. The Division of Perceptual Studies has identified 83 cases in western medical literature and has collected additional unpublished contemporary accounts wherein patients recovered complete consciousness just before death. It appears as though the damaged brain released its grip on a patient's mind and clarity returned as a result.

In 1844, a German psychiatrist named Julius reported that this occurred in 13 percent of patients who had died in his institution. In a recent investigation of end of life experiences in the United Kingdom, 70 percent of caregivers in nursing homes reported that they had observed patients suffering from dementia and confusion becoming completely lucid in their last hours before death. In a case Dr. Greyson himself investigated, a 42-year-old man developed a malignant brain tumor that rapidly grew in size. He quickly became bedridden, blind in one eye, unable to communicate, incoherent and bizarre in this behavior. He appeared unable to make any sense of his surroundings, and when members of his family touched him, he would slap as through being annoyed by an insect. He eventually stopped sleeping and would talk deliriously throughout the night making no sense. After several weeks of this, he suddenly appeared calm and began speaking coherently. He then slept peacefully. The following morning, he remained completely clear and talked with his wife, discussing his imminent death for the first time. He then stopped speaking and died.

There is no known physiological mechanism to explain this phenomenon. It is rare, but the fact that it happens has no explanation in terms of how the brain functions. It suggests the link between consciousness and the brain is more complex that most scientists think. It is as though the damaged brain prevents the person from communicating, but when the brain finally begins to die, consciousness is released from the degenerating brain.

Complex consciousness among people who have minimal brain tissue

Another phenomenon is the presence of normal or even high intelligence in people who have very little brain tissue. There are rare but surprising cases of people who seem to function normally, with normal intelligence, and normal social function, despite having virtually no brain at all. In one case, published in 2007, a high school honor student who had been accepted for enrollment by Smith College underwent surgery after she was injured and knocked unconscious in an automobile accident. An x-ray of her head just before surgery revealed that she had no cerebral cortex at all. She had just a brainstem inside her skull. When the surgeon opened her skull to operate that is exactly what he found—a brainstem and that's all.

Neurologists tell us the brainstem relays motor and sensory signals to the cerebellum and the spinal cord and integrates heart function, breathing, wakefulness, and animal functions. They also tell us the brainstem does not have the connections to perform higher cognitive functions such as thinking, perceiving, making decisions, and so forth. According to scientific knowledge as it now stands, this college-bound honor student should not have been able to formulate a thought of any kind, let alone function at a high intellectual level.

Hers is not an isolated situation. Dr. Greyson pointed to dozens of cases of patients with hydrocephalus, wherein as much as 95 percent of a brain is incapacitated due to an excess of cerebrospinal fluid, and yet many with that level of affliction have normal and even above average intelligence.

Near Death Experiences

The near death experiences [NDEs] Dr. Greyson covered in the lecture were accounts given by people who had been clinically dead for a short time and then resuscitated or revived spontaneously. He said they typically have memories of vivid sensory imagery, and an extremely clear memory of what they experienced. They often describe what they experienced as seeming "more real" than their everyday life. All of this occurs under conditions of drastically altered brain function under which the materialist model would say is absolutely impossible. Such memories are reported by between ten and twenty percent of those who are revived from clinical death. Dr. Greyson has personally investigated almost one thousand cases.

The average age at the time of the near death in these cases was 31 years, but there was a very wide range. A young girl reported an experience she'd had at eight months old while undergoing kidney surgery. The oldest to experience near death Dr. Greyson has studied was 81 at the time of his heart attack. About one third of the NDEs occurred during surgical operations, a quarter during serious illness, and another quarter as a result of life-threatening accidents. The common features of NDEs can be categorized as changes in thinking, changes in emotional state, as well as paranormal and otherworldly features.

Changes in thinking include a sense of time being altered. Often people report that time stopped or ceased to exist. The change in thinking phenomenon also included a sudden revelation or change in understanding in which everything in the universe suddenly became crystal clear. There was a sense of the person's thoughts going much faster and being much clearer than usual. Finally, there was a life review—a panoramic memory in which the person's life seemed to flash before him or her.

Typical emotions reported included an overwhelming sense of peace and wellbeing, a sense of cosmic unity and of being one with everything, a feeling of complete joy, and a sense of being loved unconditionally.

The paranormal features included a sense of leaving the physical body, sometimes called an out of body experience [OBE], a sense of physical senses such as seeing and hearing becoming more vivid than ever before. Sometimes people report seeing colors and hearing sounds that do not exist in this life, and a sense of extrasensory perception, i.e., of knowing things beyond the normal ability of the senses, such as things that are happening at a remote location. Finally, some report having visions of the future and that they entered another, unearthly world or realm of existence.

Many report they came to a border they could not cross, a point of no return that if they had crossed they would not be able to return to life. Many also say they encountered a mystical or divine being, and some report seeing spirits and loved ones who died previously and seem to be welcoming them into another realm, or in some cases sending them back to life.

As a psychiatrist, the profound after effects of NDEs are of particular interest to Dr. Greyson. Near death survivors reliably report a consistent pattern of changes in attitudes, beliefs, and values, which do not seem to fade over time. They report overwhelmingly they are more spiritual because of their experience, that they have more compassion, a greater desire to help others, a greater appreciation for life as well as a stronger sense of meaning and purpose in life. A large majority reports they have a stronger belief that we survive death of the body and no longer fear death. About half report they have lost interest in material possessions, and many report they no longer have an interest in obtaining personal prestige, status, or in competition.

Dr. Greyson said that three features of NDEs suggest consciousness is not produced by the brain: 1) Enhanced mental function while the brain is incapacitated; 2) Accurate perceptions from outside the body, such as the ability to accurately tell doctors and nurses what they saw and heard going on in the operating room; and 3) encounters with deceased persons who convey accurate information no one else could have known, including in some instances encounters with deceased persons the NDE survivor could not have known were dead at the time.

In one case, a nine-year-old boy with meningitis had an NDE in which he saw several deceased relatives, including his sister who told him he had to return to his body. As soon as he returned from death, he told his parents—who had been at his bedside for 36 hours during his ordeal. His father became very upset because his daughter was at college in a different state and was perfectly healthy as far as the father knew. The boy insisted that his sister had sent him back and had told him she had to remain.

The father left the hospital, promising his wife he would call their daughter as soon as he got home. When he tried to call her, he learned that the college officials had been trying to contact him and his wife all night to tell them the tragic news. Their daughter had been killed in an automobile accident around midnight.

By the way, if you would like to see a video of Dr. Greyson's lecture just summarized, go to YouTube and search "Dr Bruce Greyson consciousness independent of the brain." A video of the lecture should come up at the top of the list.

Children Who Recall a Past Life

Dr. Greyson also recounted information about the Division of Perceptual Studies' research into children's memories of past lives. Researchers at the University of Virginia have been conducting these investigations for more than fifty years and as a result have in excess of 2500 cases in their files. I was quite familiar with this even before I saw Dr. Greyson's lecture because of research I had done for my book, *REINCARNATION: Good News for Open Minded Christians and Other Truth Seekers*. I have in fact twice interviewed one of the Perceptual Division's key researchers who has written two books on the Division's reincarnation research findings, Jim B. Tucker, M.D., a child psychiatrist.

Anyone with an open mind who looks into what has been found will find it difficult to refute that reincarnation can and does happen. To give you a taste, I will relate a fascinating case history I also reported on in the book just mentioned. This true story began on the First of May 2000.

Imagine you and your wife [or husband] are sound asleep. Your two-year-old son James is in his crib, asleep in the next room. Suddenly you are jarred awake.

You hear your son scream, "Plane on fire! Airplane crash!"

You rush into his room, and there he is on the bed, writhing the grip of horror, kicking and clawing at the covers as if he is trying to kick his way out of a coffin.

Over and over again, your child screams, "Plane on fire! Little man can't get out!"

What happened that night was not a single occurrence. Traumatic nightly scenes like it became the norm. The nightmares became even more terrifying, and James started screaming the name of the "little man" who couldn't get out of the plane. It was "James," his own name. Other words he spoke out loud included: "Jack Larsen," "Natoma" and "Corsair."

James' father, Bruce Leininger, could not think of what to do. Eventually, in attempt to find an answer to his son's troubled nights, he embarked on a research project, armed only with the names and words his son had been shouting while in a disturbed sleep.

A devote Christian, the answer Bruce found was not the one he wanted. He came to believe his son James was the reincarnation of a World War Two fighter pilot whose plane had been hit and downed by antiaircraft fire—a pilot named James Huston who had died in 1945 after his plane suffered a direct hit and crashed.

James' mother, however, was the first to suspect the truth. At the time, James was having five nightmares a week, and his mother, Andrea, was worried. At a toy shop, Andrea and James were looking at model planes.

"Look," Andrea said. "There's a bomb on the bottom of that one."

"That's not a bomb, Mommy," James said. "That's a drop tank."

The child was two years old. How could he possibly have known about the gas tank used by aircraft in World War Two to extend their range?

As the nightmares continued, Andrea asked, "Who is the 'little man'?"

"Me," he answered.

Bruce asked, "What happened to your plane?"

"It crashed on fire."

"Why did your plane crash?"

"It got shot," James said.

"Who shot your plane?"

"The Japanese!" he said.

James said he knew it was the Japanese because of "the big red sun." He was, of course, describing the Japanese symbol of the rising sun painted on their warplanes.

Andrea began to suggest reincarnation. Wouldn't that explain it? But Bruce reacted angrily. He thought there must be a rational explanation, and reincarnation was definitely not in his mind a rational explanation.

Bruce questioned his son further. "Do you remember what kind of plane the little man flew?"

"A Corsair," two-year-old James replied without hesitation. It was a word he had shouted in his dreams.

Bruce knew a Corsair was a World War Two fighter plane.

"Where did your airplane take off?" Bruce asked.

"A boat."

"What was the name of the boat?"

James replied with certainty, "The Natoma."

Bruce did some research. He was amazed to find the Natoma Bay was a World War Two aircraft carrier. Bruce rushed to his office, where he had a dictionary of American naval fighting ships. Natoma Bay had supported the U.S. Marines' invasion of Iwo Jima in 1945.

Andrea, meanwhile, had become convinced James was reincarnated. She contacted Carol Bowman, the author of a book on reincarnation and children who remember past lives. Bowman confirmed Andrea's views, saying that the common threads were there with James, including his age when the nightmares began and his remembered death.

Bruce kept investigating. He decided to see if he could find someone named Jack Larsen, a name James had shouted repeatedly during his nightmares. Bruce was successful in finding someone who fit the time period and place. It turned out Larsen's friend James Huston had died when his plane was shot in the engine and caught fire, just as had been described by two-year-old James Leininger.

Bruce also found Huston's name on the list of 18 men killed in action on the Natoma. The discovery finally made him realize his son might actually be the reincarnation of James Huston. But he kept investigating, anyway, and everything he found served to confirm that conclusion.

One day, little James unnerved his father when he said, "I knew you would be a good daddy, that's why I picked you."

"Where did you find us?" asked an incredulous Bruce.

"In Hawaii, at the pink hotel on the beach," James said, and went on to describe his parents' fifth wedding anniversary, which had taken place five weeks before Andrea had gotten pregnant. James said that was when he "chose" the couple to bring him back into the world.

Something new emerged almost every day. On a map, James pointed out the exact location where James Huston's plane went down. Asked why he called his action figures "Billy," "Leon" and "Walter," he replied, "Because that's who met me when I got to heaven."

Eventually, the family received a phone call from a veteran who had seen Huston's plane get hit. The man had kept his knowledge to himself for more than 50 years. He described seeing the aftermath of Huston's crash on the sea below.

"He took a direct hit on the nose. All I could see were pieces falling into the bay. We pulled out of the dive and headed for open sea. I saw the place where the fighter had hit. The rings were still expanding near a huge rock at the harbor entrance."

And so it was as James had said. His plane was hit in the engine and the front exploded in a ball of flames, but that was not the end of James. He returned to this reality fifty-three years later, in 1998, with his memory intact. Perhaps he had some things here on earth he wanted to do, like flying airplanes.

How about you? Whether you come back to this world, stay in the next or move on to another, like it or not you will continue to exist. Now that you know, what is the best thing to do? Become the person you were meant to be and the person want to be. That person is someone you will probably enjoy being around, which is a good thing, because you are going to be around that person—forever.

About Stephen Hawley Martin and Other Books He Has Written

Stephen Hawley Martin is an author, ghostwriter, and publisher. You can learn more about him and get in touch with him through his website:

www.shmartin.com

If you found this book interesting, other books by Stephen you should know about are displayed on the pages that follow.

The MAGIC OF MISSION

Discover Your Purpose
Find Meaning, Achieve Success
and Above All, Be Happy

Stephen Hawley Martin

You were born with a mission. Are you pursuing it? The key to abundance is to discover your mission and pursue it, and that is what this book will help you do. It lays out a process and a formula to follow that will help you accomplish it. So, why wait? Read this book and take advantage of the magic of mission.

Kindle: ASIN: B089591VHR
PB: ISBN-13 : 979-8648691728

Bewitched?

**A Startling New Theory
of the Cause of the
Salem Witch Hysteria from
the Descendant of a Woman Hanged**

Stephen Hawley Martin

What really caused the Salem Witch Hysteria? Was it young girls pretending to be afflicted and accusing those they didn't like of causing it with witchcraft? The author doesn't think so. Don't miss this book. "Bewitched?" is a riveting, real-life murder mystery.

Kindle: ASIN: B08DXLSJNR
PB: ISBN-13: 979-8670685528

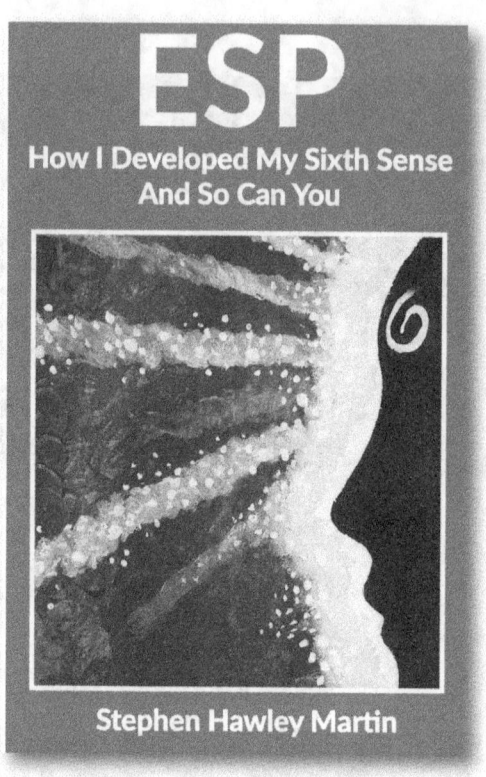

All the knowledge of the universe resides within you because at a deep level all minds, past and present, are connected. Everything that has ever happened, every thought, every idea is there. The trick is to draw out information when you need it. In this book Stephen explains how he learned to do so and how you can, too.

Kindle: ASIN: B07HHFFWP8
Paperback: ISBN-10: 1723835250

Edgar Cayce,

**The Meaning
of Life
and What to
Do about It**

Stephen Hawley Martin

You may believe humans are spiritual
beings having a physical experience,
but are you sure why we're here and
what we ought to do about it? This
book will tell this you this and much,
much more because, as the record
shows, the accuracy of information revealed by Edgar Cayce's more than
14,000 psychic readings was nothing
less than extraordinary.

Kindle: ASIN: B07L7GF3HH
Paperback: ISBN-10: 1790978114

Based on a Catastrophic Historical Event

FICTION FIRST PRIZE
WINNER
WRITER'S DIGEST

The Mt Pelée Redemption

A Thriller by
Stephen Hawley Martin

Romantic suspense at its best, this fast-paced novel won First Prize for Fiction from *Writer's Digest* and First Place for Visionary Fiction from *Independent Publisher* for good reason: It's very hard to put down. You'll be riveted as Claire flies to the island of Martinique to solve a mystery and soon realizes she's being stalked by a drug lord.

Kindle: ASIN: B00UVK8XM6
Paperback: ISBN-10: 1511675373

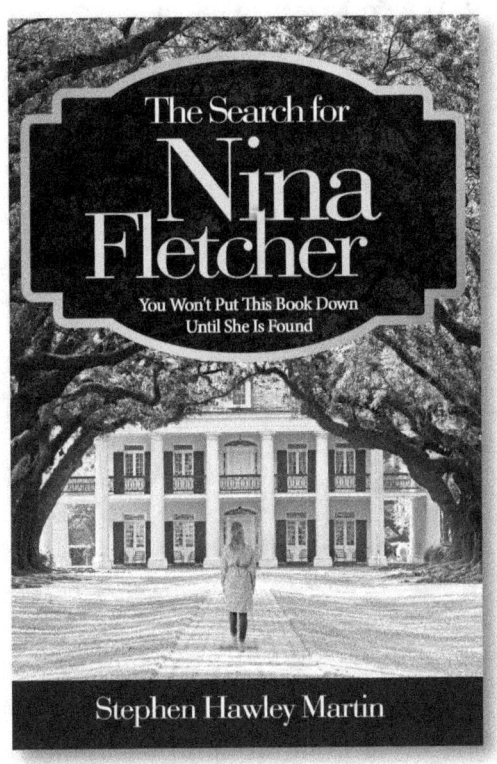

The Search for
Nina Fletcher

You Won't Put This Book Down
Until She Is Found

Stephen Hawley Martin

In this romantic suspense thriller, Rebecca wants to save the beautiful plantation home where she grew up, but to do so she must find her mother. If only she could remember what happened in the basement of the old house in Baltimore long ago. She must find out what happened there, she must!

Kindle: ASIN: B01J6MQZXS
Paperback: ISBN-10: 1535580879

Death in Advertising

FICTION FIRST PRIZE
WINNER
WRITER'S DIGEST

Stephen Hawley Martin

This whodunit set in an ad agency won First Prize for Fiction from *Writer's Digest* magazine. According to Mike Chapman, Editor-in-Chief of *ADWEEK* magazine, this novel is "A thrilling and evocative read. Masterful attention to detail brings the ad agency world to life and delivers a gripping whodunit." Get ready. You won't be able to put it down.

Kindle: ASIN: B00UIGGKUA
Paperback: ISBN-10: 1511662921

Notes